A New Approach to Quantum Theory

Series on the Foundations of Natural Science and Technology

ISSN: 2010-1961

Series Editors: C. Politis (*University of Patras, Greece*)
W. Schommers (*Forschungszentrum Karlsruhe, Germany*)
E. Meletis (*University of Texas at Arlington, USA*)

Published:

Vol. 19   *A New Approach to Quantum Theory: The Central Role of Probability*
by S. B. Sontz

Vol. 18   *Density-Of-States Function and Related Applications in*
Quantized Structures
by K. P. Ghatak and A. Biswas

Vol. 17   *Quantum Capacitance in Quantized Transistors*
by K. P. Ghatak and J. Pal

Vol. 16   *Electron Statistics in Quantum Confined Superlattices*
by K. P. Ghatak and A. Biswas

Vol. 15   *Topics in Nanoscience*
In 2 Parts
Part I: Basic Views, Complex Nanosystems: Typical Results and Future
Part II: Quantized Structures, Nanoelectronics, Thin Films
edited by W. Schommers

Vol. 14   *Elastic Constants in Heavily Doped Low Dimensional Materials*
by K. P. Ghatak and M. Mitra

Vol. 13   *Scalar Field Cosmology*
by S. Chervon, I. Fomin, V. Yurov and A. Yurov

Vol. 12   *The Origin of Natural Order: An Axiomatic Theory of Biology*
by Q. Zhao

*For further details, please visit: http://www.worldscientific.com/series/sfnst

(Continued at end of book)

Series on the Foundations of Natural Science and Technology — Vol. 19

A New Approach to Quantum Theory

Stephen Bruce Sontz

Centro de Investigación en Matemáticas, Mexico

World Scientific

NEW JERSEY · LONDON · SINGAPORE · BEIJING · SHANGHAI · HONG KONG · TAIPEI · CHENNAI · TOKYO

Published by

World Scientific Publishing Co. Pte. Ltd.

5 Toh Tuck Link, Singapore 596224

USA office: 27 Warren Street, Suite 401-402, Hackensack, NJ 07601

UK office: 57 Shelton Street, Covent Garden, London WC2H 9HE

Library of Congress Control Number: 2025035026

British Library Cataloguing-in-Publication Data
A catalogue record for this book is available from the British Library.

Series on the Foundations of Natural Science and Technology — Vol. 19
A NEW APPROACH TO QUANTUM THEORY
The Central Role of Probability

ISBN 978-981-98-2012-2 (hardcover)
ISBN 978-981-98-2013-9 (ebook for institutions)
ISBN 978-981-98-2014-6 (ebook for individuals)

For any available supplementary material, please visit
https://www.worldscientific.com/worldscibooks/10.1142/14495#t=suppl

Desk Editors: Aanand Jayaraman/Joseph Ang

Typeset by Stallion Press
Email: enquiries@stallionpress.com

A Lilia por tu amor, por tu apoyo, por todo

Abstract

...and what is the use of a book
without pictures or conversation?

Alice in *Alice's Adventures in Wonderland*
— Lewis Carroll

First, axioms of quantum theory provide a sound mathematical context
for defining and computing quantum probabilities of quantum events.
These include the conditional and consecutive probabilities of *multiple*
events as well as the probabilities of single events, although the latter are
irrelevant since science is based on correlations. These probabilities are
then to be compared with relative frequencies of *sequences* of observations.
Next, distinct models for these axioms are presented, and the concept of
isomorphism of models is defined. The only fundamental time evolution
of these probabilities is given by a new axiom called the Generalized
Born's Rule, which is not a differential equation, but involves auxiliary
formulas that depend on the model being used. Schrödinger's equation in
an auxiliary role is valid in one model of the axioms, usually called the
Schrödinger picture. So, everything here is consistent with conventional
textbook quantum physics. In other models, such as the *Heisenberg picture*,
the auxiliary equations are different, but the Generalized Born's Rule is the
same formula (*covariance*) and can give the same results (*invariance*). The
collapse of a state is simply a mathematical aid in computing conditional
probabilities, which are fundamental objects and can be computed without
this aid. Collapse is *not* a quantum event. Also, entanglement is given a new
definition in terms of quantum probability, nothing else. The *leitmotif* of
this treatise is the central importance of the probability of multiple events.
However, this approach is not a complete rendition of all of quantum theory,
even though it opens up new perspectives.

vii

Preface

Every species has come into existence
co-incident, both in time and space,
with a pre-existing closely allied species.

— Alfred Russel Wallace

In this book, I want to explain clearly, but also in the simplest possible way, the logical, axiomatic structure of quantum physics. It seems to me that the importance in quantum probability of conditional and consecutive probability is not sufficiently appreciated in the scientific community, given that these clarify the seemingly paradoxical roles of collapse and entanglement. I will be telling a new story, but some aspects of it are familiar; this is in resonance with the quote above. In short, this is a new organization of standard material with probability placed front and center. However, the history of how this physics developed, usually deemed to be highly significant pedagogically, seems to me to hide the essential structure of quantum theory. So, references to history will be few and far between.

I understand quantum probability to be a new *Generalized Born's Rule*, the rule for computing all the probabilities in quantum theory. On the one hand this rule includes the usual Born's Rule, Wigner's Rule and Lüders Rule, and on the other hand it is the fundamental time evolution equation of quantum theory. Schrödinger's equation still is valid, though only in one model of the axioms, usually known as the *Schrödinger picture*. That is the essence of this treatise, which is a new approach to standard quantum theory.

Also note that this is not an 'interpretation' of quantum theory, as that word is often used, since 'interpretations' are neither falsifiable nor verifiable. Rather this is a scientific treatise. However, I will touch on some issues that are considered in an 'interpretation' of quantum theory. This may not solve philosophical problems to everyone's satisfaction, but it is a way of throwing a bit of light on some quantum curiosities by using just three basic theoretical structures: events, states and probabilities. Nor is it a theory of everything. Other problems will, of course, require other structures.

My intended audience consists of professional scientists, philosophers and graduate students with lots of math and physics knowledge, much more than can be found in an introductory text such as [39]. Undergraduate students in physics or mathematics might find this material to be quite challenging, though they should not be discouraged from trying. Preferably, the reader will have some previous knowledge of quantum physics and Hilbert space theory. As for the latter a knowledge of functional analysis up to the spectral theory for densely defined, self-adjoint operators would be very helpful. And this will be a stand-alone quantum story, meaning that references to classical physics will be incidental, not part of the main story. It is not clear whether those with some knowledge of classical mechanics have an advantage or an impediment. Some easy pedagogical examples, which are illustrative of the general theory, will also be given. But mostly, the presentation will be quite general. The experts will note, however, that I limit myself to examples that are Type I von Neumann algebras.

In this treatise only the basics are considered. In particular, this is not a complete exposition of known quantum theory, since that would require many volumes. However, I am considering quantum theory consistent with Galilean relativity and at the same time quantum theory consistent with special relativity. If things work out that way, I am even considering quantum theory consistent with general relativity. More axioms can be introduced to cover these cases. As an analogy, note that the choice of a Hamiltonian for a specific quantum system is an additional ingredient not included in the usual axiomatic presentation.

This treatise is organized as follows. The first chapter states and then discusses the Axioms. These are standard except for Axiom 5, which is the first version of the Generalized Born's Rule seen as a time evolution equation. Next the concept of models of these axioms is introduced, including the important definition of isomorphism of models.

Chapter 2 is a long introduction to quantum probability. It begins with some comments on what a probability theory should do and proceeds to show how quantum probability does that. Here, the important topics of conditional and consecutive probability are discussed. This is not new, neither in theory nor experiment, since one has to assign theoretical probabilities to sequences of multiple events and only relative frequencies of such sequences of events are measured experimentally. Actually, the theoretical probability of single events is an irrelevant mathematical nicety, since science is based on the correlations of multiple events.[1] Then quantum probability is compared and contrasted with its well known intellectual competitor: classical probability as a topic in measure theory. But I do not consider comparisons with Bell's inequalities, hidden variables, classical physics and other such side issues, because my focus is on quantum theory as a new type of probability theory and nothing else. Also, there is a description of the spectral theorem in terms of events, rather than in terms of eigenvalues and eigenvectors, which anyway is only a special case.

Next Chapter 3 applies this theory to Entanglement. Then in Chapters 4 to 8 I discuss various issues that have filled volumes for decades. These are Schrödinger's Cat, measurement problems, the EPR paper, determinism and probability and philosophical questions. Nonetheless, these recurrent issues are not clarified by the so-called basic equation of quantum theory, Schrödinger's equation. While I may not resolve any of these issues to the reader's satisfaction, I do try to clarify what they are (or should be) about in terms of quantum events, quantum states and quantum probability. In Chapter 9, I discuss four problems posed by Isham in [21].

In the rest of the treatise I consider topics that strictly speaking have little or nothing to do with quantum theory. I include them since it is an appropriate moment for commenting on them. In Chapter 10, I attempt to dismiss 'interpretation', in the contemporary sense of that word in reference to quantum theory, from all scientific consideration. Chapter 11 is devoted to the 'wave function' about which I expect the reader to know a lot already. I include it in order to rid quantum theory of residual classical language that slips into discussions of it. In the final Chapter 12, I put forward an unlikely proposal and bid my kind reader farewell.

Readers with a lot of background can get the basic gist of this treatise by reading the sections on Axiom 5 and quantum conditional probability and then skipping forward to the chapter on Entanglement. I would like to point out that I originally thought that Entanglement could not be explained without using the collapse of the state associated with a measurement. So, I examined this case carefully in order to understand the fundamental role of the collapse condition. What I found out surprised me; an analysis in terms of the conditional probability of sequences of quantum events suffices.

---

[1]This is also in resonance with the quotation of Wallace.

A central point of this treatise is that theoretical science provides us with quantitative correlations, which are then tested by observations and experiments. In particular, the correlations of sequences of multiple events form the basis of quantum theory as expressed in the Generalized Born's Rule. It will be argued that correlation alone can be incorrectly inferred from experience, and this is quite so. This is known colloquially as the sauce Béarnaise syndrome and could well be the origin of much superstition. And this is exactly why ever more careful observations and experiments are also critical elements of scientific activity. But here you will only see a theoretical treatise, which as with all such works requires empirical verification. In many ways this treatise is prologue.

Scientific activity also requires a combination of curiosity and narrative. While some sort of scientific activity in this sense is found in all cultures, the contemporary version of science is not a necessary aspect of all human cultures. This treatise is a narrative driven by my curiosity to understand for myself as well as by my desire to explain to the scientific community the basic structure of quantum mechanics. The contemporary concept of science is not an invention of the modern world, even though it is a human invention of an intellectual sort. One can learn how science, essentially as it is understood nowadays, was practiced in Hellenistic society beginning around 300 B.C. from the fascinating book [35] by L. Russo. As Russo points out, some 'modern' scientists do not grasp completely what Archimedes and Co. were doing in Ancient Antiquity. One necessary scientific activity, then as now, is the construction of mathematical models which relate to observations. It is a two-way street, but usually the flow starts on the observational side. In this treatise the historical process of empirical discovery is not my interest. Rather I will directly jump over to the theoretical side by starting with an axiomatization of quantum theory. This methodology is what Maxwell's Equations are all about, even though they are not typically called Maxwell's Axioms. Maybe they should be renamed.

Many people have helped me in understanding the matters discussed here. Sometimes that help was a detailed explanation, but often it was just a casual question or comment which focused my attention on an important matter. A partial list of those whom I have known personally must include Luigi Accardi, David Brydges, Jaime Cruz Sampedro, Matthew Dawson, Micho Đurđevich, Jean-Pierre Gazeau, Leonard Gross, George Hagedorn, Brian Hall, Ira Herbst, Jim Howland, Dennis Stepanek, Larry Thomas and Carlos Villegas.

And during my undergraduate days I was advised to read the masters. So, I am indebted also to those whom I have only known through their published work. Most notable among those are Valentine Bargmann, Charles Darwin, Paul Dirac, Richard Feynman, David Hilbert, John von Neumann, Erwin Schrödinger, and Eugene Wigner. I thank all of these as well as many more, too numerous to recall and list their names. I also thank CIMAT for a sabbatical leave used to advance this project to near completion. Of course, any shortcomings or errors are due to my own imperfect understanding of non-intuitive[2] matters.

And finally I most warmly thank World Scientific and my marvelously helpful editor, Joseph Sebastian Ang.

Stephen Bruce Sontz
Centro de Investigación en Matemáticas, A.C.
(CIMAT)
Guanajuato, México
April 2025

---

[2]Intuition is not an objective criterion. In the current context intuition refers to my own time dependent insight and its limitations.

Clarifying the Confusion

So in physics a paradox is only
a confusion in our own understanding.

— Richard Feynman

Here is why I wrote this book. There has been a lot of confusion for those used to the standard textbook quantum theory such as found in [39] with these features:

- There is only one fundamental equation of time evolution, Schrödinger's equation, which is a differential equation. That equation continuously and deterministically describes the change in time of the sole, unique physical characteristic of a quantum system, namely its state.
- Probability enters the theory only via single measurements, which are accompanied by a second time evolution (known as collapse). This is discontinuous and probabilistic and is not described by a differential equation. So, there are two distinct time evolutions, unlike any other physical theory. This is the measurement problem of quantum theory.
- Quantum states have weird non-classical properties, including among others entanglement and collapse, which defy the intuition of most mortals who can think only classically.
- The Heisenberg and interaction pictures are not basic, but only serve as computational conveniences for dealing with problems in the basic Schrödinger picture.

The confusion — some would say outright paradox — is how to consistently deal with all this. Consequently, there is a lot of discussion about what quantum theory really means, but nothing really in the way of progress.

An outline of the way out of this confusion is given as follows:

- There is one basic time evolution equation, known as the Generalized Born's Rule. It is not a differential equation. However, Schrödinger's equation is still valid in one model of the new axioms, and so all the theory based on that equation is also valid in the context of that model.
- The Generalized Born's Rule uses states as secondary mathematical objects to calculate probabilities, which describe the most important property of *sequences* of events, namely, their relative frequencies. In no way do states characterize a quantum system, which is described better by its possible events, their probabilities as well as its state.
- The probability of single events can be calculated, but is irrelevant. The only contact with experiment is through the probabilities of two or more events. *Science is based on correlations.* Measurements are just one type of event and play no distinguished role in this approach. Collapse is just a misleading way of speaking about how to calculate conditional probability, a basic concept. Collapse is not a physical event requiring any further explanation as to 'how it happens', but rather is just one step in an algorithm for computing a conditional probability. Also, entanglement is *newly defined* as the lack of independence of events in quantum probability; it is described by explicit mathematical formulas and is not paradoxical, although perhaps not intuitive.
- Classical mechanics also describes non-intuitive situations, with many having to do with rotating bodies. But we can rely on mathematics to keep us on track with quantum 'weirdness' just as we do in the case of classical 'weirdness'.
- The various pictures of quantum theory are understood as different models of the same basic axioms. None is more basic than any other.

This original approach clears up a lot of the muddy water of standard quantum theory by emphasizing the role of sequences of events, rather than only dealing with single events. But still there does remain a major open problem concerning measurements, even though it is not the measurement problem of quantum theory (as described above), which is resolved as will be discussed in Chapters 4, 5 and 9.

There is often some confusion about the role of Planck's constant $\hbar$ in a treatise on quantum theory. But just as one can take the speed of light to be $c = 1$ in a relativistic text, so here we will take $\hbar = 1$. This convention implies that energy has the same dimensions as inverse time and that linear momentum has units of inverse length. Also angular momentum and action are dimensionless.

Only when passing to the non-relativistic limit do we put c explicitly into the formulas and then consider the limit $c \to \infty$. Similarly, only when considering the semi-classical limit do we put $\hbar$ into our formulas and then consider the limit $\hbar \to 0$.

In this context quantization of a classical mechanical theory seems as incongruous as the 'relativization' of such a theory.

Moreover, there is some confusion about how just some events and (in particular, most or all measurements) should be considered 'classical' and therefore distinguished from the quantum events of this treatise. Now, the resolution of this confusion is that *all* events and *all* measurements are viewed here the same way, namely as orthogonal projection operators. Simply put, in this theory there are no 'classical' events and no 'classical' measurements, just events and measurements.

About the Author

Stephen Bruce Sontz received a Ph.D. specializing in mathematical physics from the University of Virginia. He has published research papers about Segal–Bargmann analysis, Toeplitz and co-Toeplitz quantization, non-commutative geometry and most recently the foundations of quantum theory. Moreover, he is the author of an introductory text on quantum theory aimed at undergraduates with some background in mathematics but nothing at all in physics. Currently, he is a research professor at the Centro de Investigación en Matemáticas, A.C. in Guanajuato, Mexico.

Contents

Abstract   vii

Preface   ix

Clarifying the Confusion   xv

About the Author   xix

List of Notations and Abbreviations   xxv

Chapter 1.   Definitions and Axioms   1

   1.1   A Few Preliminaries   2
   1.2   Axiom 1: Kinematics   3
   1.3   Axiom 2: States and Events   6
   1.4   Axiom 3: Spin and Statistics   8
   1.5   Axiom 4: Time Independent Born's Rule 10
   1.6   Axiom 5: Dynamics 12
   1.7   Models of the Axioms 13
   1.8   How Many Time Evolutions? 20
   1.9   Relation to Observation 21
   1.10 An Incomplete Theory 22

Chapter 2.   Quantum Probability   23

   2.1   Introduction 23
   2.2   The Physical Assumptions 24
   2.3   The Mathematical Model 25
   2.4   The Finite Dimensional Case 29
   2.5   Quantum Probability: The First Steps 32
   2.6   Quantum Conditional Probability 35

2.7   Quantum Consecutive Probability   39
2.8   Quantum Probability of Two Events   43
2.9   Collapse as Part of an Algorithm   50
2.10   Generalized Born's Rule with a State   52
2.11   Generalized Born's Rule with no State   56
2.12   Probability Amplitudes 58
2.13   Quantum Integrals 59
2.14   Born's Rule Redux 60
2.15   Comparison with Classical Probability   62
2.16   Expected Value 66
2.17   Dynamics: The Generalized Born's Rule,
        The Final Version 67
2.18   Quantum Information 71
2.19   Afterthoughts on Events 71
2.20   The Fate of the State 73
2.21   Events Suffice for All Observables   74
2.22   The Irrelevance of Single Events   74
2.23   Preparation of the State   74
2.24   Is Standard Quantum Theory Being Changed?   76

Chapter 3.   **Entanglement**   **79**
3.1   A Standard Example   80
3.2   Entangled States   86
3.3   Collapse Is Marginalized   89
3.4   Entanglement without Tensor Products   91
3.5   Commuting vs. Non-commuting Events   94

Chapter 4.   **Schrödinger's Cat**   **95**

Chapter 5.   **Measurement Problems**   **101**

Chapter 6.   **The EPR Paper**   **107**

Chapter 7.   **Determinism and Probability**   **113**

Chapter 8.   **Philosophical Questions**   **117**

Chapter 9.   A Quaternity of Problems   **121**

Chapter 10.   Interpretation   **125**

Chapter 11.   The Wave Function   **127**

Chapter 12.   A Proposal and a Farewell   **131**

Bibliography   133

Index   137

List of Notations and Abbreviations

Symbol	Meaning
A	Self-adjoint operator
(a, b)	Open interval $\{t \in \mathbb{R} \mid a < t < b\}$ for real numbers $a < b$
B	Borel subset of $\mathbb{R}$
$\mathcal{B}$	Basis
$\mathcal{B}(\mathbb{R})$	The σ-algebra of Borel subsets of $\mathbb{R}$
$\mathrm{Bij}(A)$	The set of bijections of a set A to itself
C^∞	Having all derivatives of all orders
$\mathbb{C}$	The set of complex numbers
$\mathbb{CP}(\mathcal{H})$	The complex projective space of $\mathcal{H}$
$\dim$	Dimension
e	Base of the natural exponentials and natural logarithms
E	Quantum event or, equivalently, (orthogonal) projection
E^c	$I - E$, the complementary quantum event of E
$\mathcal{E}$	The set of all quantum events
E_t	One-parameter group of bijections of $\mathcal{E}$, where $t \in \mathbb{R}$
$\mathcal{E}(X)$	Expected value of the random variable X
$\mathcal{E}_\psi(A)$	Expected value of the operator A given the pure state represented by ψ
f	A function
H	Hamiltonian operator
H_{free}	Free Hamiltonian operator
H_{int}	Interacting Hamiltonian operator

$\mathcal{H}$	Hilbert space over the field $\mathbb{C}$
$\mathcal{H}'$	Hilbert space dual to $\mathcal{H}$ of $\mathbb{C}$-linear functionals $\mathcal{H} \to \mathbb{C}$
$(\mathcal{H}, \mathcal{V}, E, S)$	Model of quantum theory
$\hbar$	Planck's normalized constant
i	The complex unit $\sqrt{-1}$
I	Identity operator (on the appropriate vector space)
$I_{\mathcal{H}}$	Identity operator on $\mathcal{H}$
id	The identity function (on the appropriate domain)
l	Linear functional
$\mathcal{L}(\mathcal{H})$	The set $\{T : \mathcal{H} \to \mathcal{H} \,|\, T$ is linear and bounded$\}$
$L^2(\mathbb{R}^3)$	Hilbert space of equivalence classes of square integrable functions $f : \mathbb{R}^3 \to \mathbb{C}$
m_k	kth moment of a measure
$\mathcal{M}(\Omega)$	Vector space of all measurable functions $f : \Omega \to \mathbb{C}$
$\mathbb{N}$	The set of non-negative integer numbers
$\mathbb{N}^+$	The set of strictly positive integer numbers
pvm	Projection valued measure
P	Probability; projection valued measure
P_A	Projection valued measure of the self-adjoint operator A
$P_\rho(E)$	Quantum probability of the event E given the density matrix ρ
$P_\psi(E)$	Quantum probability of the event E given the pure state represented by ψ
$P_\rho(T \in B)$	Quantum probability that the observable T lies in the set B, given the density matrix ρ
$P_\psi(T \in B)$	Quantum probability that the observable T lies in the set B, given the pure state represented by ψ
$P_\rho(E_1 \,|\, E_2)$	Conditional quantum probability of the event E_1, given an earlier event E_2 and the density matrix ρ
$P_\psi(E_1 \,|\, E_2)$	Conditional quantum probability of the event E_1, given an earlier event E_2 and the pure state represented by ψ
$P_\rho(E_1, E_2)$	Consecutive quantum probability of the event E_1 and then later the event E_2, given the density matrix ρ

$P_\psi(E_1, E_2)$	Consecutive quantum probability of the event E_1 and then later the event E_2, given the pure state represented by ψ
$\mathbb{R}$	The set of real numbers
$\mathrm{Res}(A)$	Resolvent set of the operator A
$\mathcal{S}$	The set of all quantum states
S_t	One-parameter group of bijections of $\mathcal{S}$ where $t \in \mathbb{R}$
S_1, S_2, S_3	2×2 spin matrices
S^2	2×2 total spin matrix
$\mathrm{Spec}(A)$	The spectrum of the operator A
$SU(2)$	The special unitary Lie group of 2×2 complex matrices
$\mathrm{Supp}\, P$	Support of a pvm P
t	Time
T	Linear operator
Tr	Trace of a trace class operator
$\mathcal{T}(\mathcal{H})$	Banach space of trace class operators $T : \mathcal{H} \to \mathcal{H}$
U	Unitary operator; open set
$U(s, t)$	Time evolution operator
$\mathcal{V}$	A von Neumann algebra
V_j	Spectral subspace
X	Classical random variable
$[0, 1]$	Closed interval $\{ t \in \mathbb{R} \mid 0 \le t \le 1 \}$
δ_{ij}	Kronecker delta function
$\varepsilon_1, \varepsilon_2$	Standard orthonormal basis of $\mathbb{C}^2$
λ	Real or complex number
$\{\lambda\}$	The set whose only element is λ
Λ	The empty sequence of events
μ	Measure
μ_X	Distribution of the random variable X
ρ	Density matrix
σ	Standard deviation
σ_k	kth central moment of a measure
ϕ	Unit vector in $\mathcal{H}$; representative of a pure state
ϕ_t	Representative of a pure state that depends on time $t \in \mathbb{R}$

$\lvert\phi\rangle\langle\phi\rvert$	Dirac notation for the rank 1 density matrix associated with a pure state represented by ϕ
$\langle\phi\rvert$	Dirac bra — a notation for the linear functional in $\mathcal{H}'$ associated to the unit vector ϕ in $\mathcal{H}$
$\lvert\phi\rangle$	Dirac ket — an alternative notation for the vector ϕ in $\mathcal{H}$
χ_S	The characteristic function of the set S
ψ	Unit vector in $\mathcal{H}$; representative of a pure state
ψ_t	Representative of a pure state that depends on time $t \in \mathbb{R}$
Ω	Classical probability space; region in Minkowski space-time
$(\Omega, \mathcal{F}, P)$	Classical probability space, its σ-algebra and its probability function
$\emptyset$	The empty set
$\langle\cdot,\cdot\rangle$	Inner product
$\lVert\cdot\rVert$	Norm
$\lVert\cdot\rVert_{op}$	Operator norm
$\cap, \bigcap$	Intersection of sets
$\cup, \bigcup$	Union of sets
$\wedge, \bigwedge$	Infimum of events or lattice elements
$\vee, \bigvee$	Supremum of events or lattice elements
$\otimes$	Tensor product

Chapter 1

Definitions and Axioms

I had been told that Euclid
proved things and was much
disappointed that he started
with axioms.

— Bertrand Russell

In [39], I put the axioms in the penultimate chapter of the book. This
was a pedagogical choice. The intended audience for that book consists of
people with no prior background in physics, and so I wanted to present
first the particulars and get to the logic behind it all later. In fact, since
many novices are allergic to axioms and logic, I even made that chapter
optional. But this treatise addresses a much more advanced audience.
I not only expect your interest in these details, but also your understanding
of their import. One point is that there is a way to associate certain
aspects of this mathematical theory with physical phenomena. A theory,
any theory, is interesting and important if there are a sufficient number
of physical phenomena that are adequately described by it. There is no
need to claim that this theory, or any theory, is adequate for describing
all possible physical phenomena. So, here is the sense in which I use the
expressions 'physical characteristic' or 'observable quantity', for example.
These are like 'point' and 'line' in geometry; they are undefined expressions
subject to possible correspondence with certain physical phenomena. And
in analogy to geometry, where 'point' and 'line' do not describe all spatial
phenomena, these definitions and axioms are not intended to describe all
physical phenomena in terms of these basic concepts: events, states and
probability, which will be defined in terms of Hilbert space structures. And
good definitions are often non-trivial.

These axioms are intended to provide a starting point for understanding
standard 'textbook' quantum mechanics as used on a daily basis by

scientists and engineers. If you prefer to start with other axioms that have these as logical consequences, then you are implicitly accepting the rest of this treatise, provided that I have made no mistakes. If you prefer to start with other axioms that contradict these, then the points of discrepancy should be subjected to experimental tests. These axioms are not intended to be complete, but are meant to give an explicit, logical basis for the rest of the treatise. Nor are they intended to be the final, most efficient way to give an axiomatization of quantum theory. These axioms are not the same as those given in [39]. Neither of these is intended to be a complete list of axioms. The statement of each axiom terminates with the symbol.   ∎

I will use the spectral theorem for self-adjoint, even possibly unbounded, operators acting in a complex Hilbert space. However, this is more than saying that every such operator has a representation (or diagonalization) in terms of eigenvalues with respect to some orthonormal basis of eigenvectors. Anyway, that is only one special case of the spectral theorem. The general statement, which we will use, is that such operators are in one-to-one and onto correspondence with projection-valued measures (pvm) defined on the Borel subsets of $\mathbb{R}$. The importance of this result is that the pvm itself contains all the information of the corresponding operator. Most importantly, since projections are events in quantum theory, this gives a family of events associated to the operator. Moreover, those events under certain conditions can be identified as measurements. And that is how measurement is treated in this theory. More mathematical details are found in Section 2.3.

There is a sort of critique of formal rules and axioms that objects to the lack of prior justification of the rules. That they work well for all observations is not accepted as a satisfactory answer. Of course, any proposed explanation of 'why' the rule is correct is itself subject to the same criticism. And so on *ad infinitum*. Such criticism could be applied to the Generalized Born's Rule, to which I respond: *Hypotheses non fingo*.

1.1   A Few Preliminaries

I try to use standard notations and definitions. (For example, see [42].) Throughout this treatise *self-adjoint operators* are understood to be densely defined and possibly unbounded.

We let $\mathbb{C}$ denote the field of complex numbers and $i = \sqrt{-1} \in \mathbb{C}$. We also define $\mathcal{L}(\mathcal{H}) := \{T : \mathcal{H} \to \mathcal{H} \,|\, T \text{ is linear and bounded}\}$, where $\mathcal{H}$ always denotes a complex Hilbert space, usually assumed to be separable.

We note that $\mathcal{L}(\mathcal{H})$ when equipped with the operator norm, denoted by $||\cdot||_{op}$, and the adjoint operation, denoted by $T \mapsto T^*$, is both a C^*-algebra as well as a von Neumann algebra. The inner product on $\mathcal{H}$, denoted as $\langle \cdot, \cdot \rangle$, is anti-linear in the first entry and linear in the second. The norm on $\mathcal{H}$ is denoted as $||\cdot||$.

Here is some Dirac notation we will use. Every element in the dual Hilbert space $\mathcal{H}' := \{l : \mathcal{H} \to \mathbb{C} \mid l \text{ is linear and bounded}\}$, according to the Riesz representation theorem, can be written for a unique $\phi \in \mathcal{H}$ as $l = \langle \phi |$, where the *bra* $\langle \phi |$ is defined by $\langle \phi | \psi := \langle \phi, \psi \rangle$ for all $\psi \in \mathcal{H}$. Every vector $\psi \in \mathcal{H}$, a Hilbert space, can also be denoted as a *ket* $|\psi\rangle$. Then the definition of bra becomes $\langle \phi | |\psi\rangle := \langle \phi, \psi \rangle$. For a pair of vectors $\phi, \psi \in \mathcal{H}$, we define

$$|\psi\rangle\langle\phi| := |\psi\rangle \otimes \langle\phi| \in \mathcal{H} \otimes \mathcal{H}'.$$

This is then identified with the element in $\mathcal{L}(\mathcal{H})$ defined for all $\alpha \in \mathcal{H}$ by

$$|\psi\rangle\langle\phi| \, \alpha := \langle\phi, \alpha\rangle \psi.$$

Its operator norm satisfies $|| \, |\psi\rangle\langle\phi| \, ||_{op} = ||\psi|| \, ||\phi||$, while its operator adjoint satisfies $(|\psi\rangle\langle\phi|)^* = |\phi\rangle\langle\psi|$. If both ψ and ϕ are also non-zero, then $|\psi\rangle\langle\phi|$ is a rank 1 operator. If ϕ is a unit vector, then $|\phi\rangle\langle\phi|$ is both a projection and a self-adjoint, positive, trace 1 operator, that is, a *density matrix*.

After the Contents a list of standard notations and abbreviations from physics and mathematics can be found.

1.2 Axiom 1: Kinematics

Associated to every quantum system there is a non-commutative von Neumann algebra $\mathcal{V}$ in $\mathcal{L}(\mathcal{H})$. The projections in $\mathcal{V}$ are called *(quantum) events* and are associated with the physical events of the system. There is also a set $\mathcal{O}$ of *observables* which are (possibly densely defined, unbounded) self-adjoint operators, each of which is associated with a physical observable of the system. In particular, every event in $\mathcal{V}$ is in $\mathcal{O}$. Also, each self-adjoint operator A acting in $\mathcal{H}$ that is in $\mathcal{O}$ necessarily is *affiliated*[1] to $\mathcal{V}$, which means that its projection valued measure (pvm) P_A (see Definition 2.3.1) satisfies $P_A(B) \in \mathcal{V}$ for all Borel subsets B of $\mathbb{R}$. ∎

---

[1]This is motivated by physics, since the events $P_A(B)$, namely, that A takes a value in B, should be observables if A is observable.

A von Neumann algebra is a sub-$*$-algebra $\mathcal{V}$ of $\mathcal{L}(\mathcal{H})$ such that $\mathcal{V} = \mathcal{V}''$, the double commutant. There might be physical reasons for requiring $\mathcal{V}$ to be a factor. The first condition in the axiom implies that $\dim_{\mathbb{C}} \mathcal{H} \geq 2$. This axiom does not assert that every observable quantity of a physical entity has a corresponding self-adjoint operator in quantum theory. The most important example of such an observable is the time of a physical event. The criterion of non-commutativity of $\mathcal{V}$ reflects Dirac's often stated opinion that the essential characteristic of quantum theory is that the observables do not commute.

Since this is a presentation based on events and in particular events of the form $P_A(B)$ (using the notation of the axiom), it is more convenient, almost obligatory, to use von Neumann algebras instead of the more general structure of C^*-algebras. The point is that if one starts with a C^*-algebra $\mathcal{C}$ and one has a self-adjoint $A \in \mathcal{C}$, then the associated events $P_A(B)$ do not necessarily lie in $\mathcal{C}$, though they do lie in the smallest von Neumann algebra containing $\mathcal{C}$.

The mathematical definition of a quantum event as a projection operator must be considered with care, since it does not correspond exactly to the word 'event' in common English usage. A projection operator $E \notin \{0, I\}$ has spectrum $\{0, 1\}$. Sometimes it is said that a quantum event is a YES-NO phenomenon. In other words, a quantum event has exactly two possible eigenvalues. But a physical event colloquially means that only one thing has occurred. For example, suppose there is one beta unstable, radioactive nucleus in an atomic trap. Suppose that it decays in a certain time period. Common usage has it that beta decay occurred. But if it does not decay in that time period, common usage has it that nothing happened, that there was no physical event. But the second alternative is just the NO or 0 eigenvalue of the event of beta decay in that time interval. In either case, we have the *same* quantum event, but with two distinct values. For example, we say that $P_A(B)$ is the quantum event that the observable A takes a value in B. But in general this quantum event has *two* eigenvalues. Just as with any other non-trivial self-adjoint operator, a non-trivial quantum event can represent more than one value. We can also think of the eigenvalue 1 of the event $P_A(B)$ as meaning that a measurement associated with A produced a value in B, while the eigenvalue 0 says that the same measurement produced a value in $\mathbb{R} \setminus B$, the set complement of B in the real line. But measurements do not play a distinguished role in this treatise; they are just events $P_A(B)$, where B is a small interval in $\mathbb{R}$. A nuclear reaction in a star in a distant galaxy is an event and so is an atomic transition measured in a laboratory across town. And events do play a central role here.

This definition of an event as a projection operator may seem strange, even though it is found in many places in the literature[2]. So, here are some further comments. First, events E are physical observables and therefore should be represented by self-adjoint operators. So, $E = E^*$. But events should have only two values, one corresponding to an occurrence and the other to a non-occurrence. With no loss in generality we can take these values to be 0 and 1. It is a now an innocuous convention to assign 1 to occurrence and 0 to the opposite. This narrows the spectrum to lie in the two point set $\{0, 1\}$. Now, a real number λ satisfies $\lambda^2 = \lambda$ if and only if $\lambda \in \{0, 1\}$. This motivates requiring $E^2 = E$. Putting these two conditions on E together, we get that E is an orthogonal projection operator.

Of course, if one thinks deterministically, an event is something that must happen or, on the contrary, must not happen. That is not a description of quantum events, which are merely possible happenings. Perhaps, thinking about the meaning of the French word *eventuellement* (which translates into English as *possibly*) may help the reader.

It is traditional to speak in quantum theory in terms of the self-adjoint operators acting in a given Hilbert space $\mathcal{H}$. This is not logically necessary, since in functional analysis one proves that these mathematical structures are in bijective correspondence with two other structures. These structures are on the one hand projection valued measures (pvm's) with values in $\mathcal{L}(\mathcal{H})$ defined on the Borel σ-algebra of $\mathbb{R}$ and on the other hand strongly continuous unitary groups acting on $\mathcal{H}$. For example, one can define the commutativity of two self-adjoint operators S and T by one of these equivalent definitions:

- Let P_S and P_T, denote the pvm's of S and T, respectively. Then we say that *S and T commute* if $P_S(B)$ commutes with $P_T(C)$ for all Borel subsets B, C of $\mathbb{R}$. (Note that each of the families $\{P_S(B) \,|\, B \in \mathcal{B}(\mathbb{R})\}$ and $\{P_T(C) \,|\, C \in \mathcal{B}(\mathbb{R})\}$ is commutative. This definition requires that their union is also a commutative family.)
- Let e^{irS} and e^{isT} for $r, s \in \mathbb{R}$ denote the unitary groups of S and T, respectively. Then we say that *S and T commute* if e^{irS} commutes with

---

[2]However, I do not know who was the first to give this definition.

e^{isT} for all $r, s \in \mathbb{R}$. (Note that here as well each of the families of unitary operators $\{e^{irS} \mid r \in \mathbb{R}\}$ and $\{e^{isT} \mid s \in \mathbb{R}\}$ is commutative. This definition requires that their union is also a commutative family.)

The next axiom is really just a continuation of Axiom 1. It is included as a separate axiom because of tradition.

1.3   Axiom 2: States and Events

Every quantum system is described by probabilities that are computed by using its set of events (see Axiom 1) and (some subset of) the states of its von Neumann algebra $\mathcal{V}$. ∎

In this treatise we will consider all states of the von Neumann algebra, though in other formulations selection rules are used to restrict to a subset of these. Unfortunately, the terminology 'state' is so widely used that there is no hope of ever replacing it with a neutral term. It is a term that drips with meanings from classical physics as well as from everyday life. But in quantum theory as presented here it is a mathematical term with a mathematical definition, given below, which clearly says a state is associated with a quantum system and not with any subsystem or particle composing that system. We shall see the import of this remark when considering entanglement in Chapter 3.

In the following, we will only have examples in the case when the von Neumann algebra $\mathcal{V} = \mathcal{L}(\mathcal{H})$, since many quantum systems are covered by this case. In introductory texts it is the only case considered, since that already puts a lot of mathematical burden on beginners. I did this myself in [39]. While this is the only case familiar to many physicists, there are other von Neumann algebras used in quantum theory. When a physicist speaks of a Hilbert space as being the setting for the discussion of a quantum system, the underlying, usually implicit assumption is that the appropriate von Neumann algebra is $\mathcal{L}(\mathcal{H})$. But that could be a *Type I error*. The problem is to find the correct von Neumann algebra, and not only the correct Hilbert space as is often thought. This is just a part of the problem of *quantization*. An important point here is that the lattice of events is central to quantum theory, and its structure depends on the von Neumann algebra being used. Note that the quantization problem is not a problem within quantum theory itself, but rather a problem of how to arrive at a quantum theory for a particular system or, even more difficult, for some class of systems. This non-trivial and important problem will not be addressed in this treatise.

Definition 1.3.1. A (*normal*) *state* is a linear map $l : \mathcal{V} \to \mathbb{C}$ satisfying

- (Positivity Preservation)
  $l(T) \geq 0$ for all $T \in \mathcal{V}$ that are *positive* (meaning that $T = T^* \geq 0$).
- (Normalization)
  $l(I) = 1$, where $I \in \mathcal{V}$ denotes the identity operator.
- (Continuity)
  l is σ-weakly continuous.

This is (or is equivalent to) the standard definition of normal state in operator theory. It is a theorem (see [6]) that for any normal state l there exists a *density matrix* $\rho \in \mathcal{L}(\mathcal{H})$ (that is, a self-adjoint, positive, trace class operator with $Tr\,\rho = 1$) such that $l(T) = Tr(T\rho)$ for all $T \in \mathcal{V}$. Here, Tr is the trace of a trace class operator. Thus, $\rho \in \mathcal{T}(\mathcal{H})$, the Banach space of trace class operators, which is the pre-dual to the Banach space $\mathcal{L}(\mathcal{H})$. It is important to note that ρ is not necessarily in $\mathcal{V}$. In this case we let l_ρ denote this normal state and we say that l_ρ is a *mixed state*. Conversely, for every density matrix $\rho \in \mathcal{L}(\mathcal{H})$, the linear map l_ρ is a normal state. In this treatise, this will be the only sort of state considered and so the qualifier 'normal' will be omitted from now on. It turns out that every state l is a bounded linear functional with operator norm $||l||_{op} \leq 1$. We let $\mathcal{S}$ denote the set of all states.

A consequence of this axiom is that every *unit vector* $\psi \in \mathcal{H}$, that is $||\psi|| = 1$, determines a state l_ψ defined by $l_\psi(T) := Tr(|\psi\rangle\langle\psi|T) = \langle\psi, T\psi\rangle$ for all $T \in \mathcal{V}$. Moreover, the state l_ψ does *not* uniquely determine a unit vector that defines it, since $l_\psi = l_\varphi$ for all unit vectors φ that satisfy $\varphi = \lambda\psi$ for some $\lambda \in \mathbb{C}$. (Necessarily, we have $|\lambda| = 1$.) Such a state l_ψ is called a *pure state*. We say that the non-unique unit vector ψ *represents* the pure state l_ψ. Colloquially, one says that a unit vector *is* a pure state or simply a state, even though this is mathematically incorrect. Both l_ψ and the unit vector ψ exist in the mathematical sense of existence, but only the former is called the pure state. This notion of existence is not to be confused with ontology.

Since $\dim_{\mathbb{C}} \mathcal{H} \geq 2$, it is easy to construct mixed states which are not pure states. In common parlance, especially among physicists, the expression 'mixed state' is often reserved for those states l_ρ which are not pure states. This is due in part to the emphasis given to pure states in many applications. Indeed, often one only considers the pure states, which are usually simply called *states*. It is important to note that the set of pure states is in bijective correspondence (by the mapping $l_\psi \mapsto \mathbb{C}\psi$, where ψ is a unit vector) with the set of all one-dimensional subspaces of the Hilbert space $\mathcal{H}$. The latter set is called the *(complex) projective space* associated to $\mathcal{H}$ and is denoted as $\mathbb{CP}(\mathcal{H})$. This projective space has many interesting mathematical properties.

The set of quantum events should be in bijection with the physical events of a given quantum system. If not, then one has not correctly chosen the von Neumann algebra for the system. The choice of the self-adjoint operators, which are physically relevant, is not so obvious. It is generally accepted that the self-adjoint operators should be affiliated with the von Neumann algebra, as specified in Axiom 1, though it may be the case that not all affiliated self-adjoint operators will have a physical significance. As for the states, physical considerations could indicate that they should form some proper subset of all the states of the von Neumann algebra. To my way of thinking, this is not an *ad hoc* super-selection rule, but just a way of carefully choosing an accurate probabilistic model for a quantum system. Such a choice of model is called a *quantization*. My rule of thumb is that a quantization should be based on a von Neumann algebra that is generated by the smallest set of events with the smallest set of states needed to understand the quantum system. However, quantization remains to this day as much an art as a science.

1.4   Axiom 3: Spin and Statistics

The most basic quantum entities are either *bosons*, all of which have integer spin, or *fermions*, all of which have half-integer spin. All other quantum entities are composites of these.   ∎

The Hilbert space $\mathcal{H}$ for a single boson or a single fermion carries a *unitary irreducible representation* of the *Lie group* that is the universal cover[3] of the rotation group $SO(3)$ of $\mathbb{R}^3$. That representation defines the value of the spin. It seems that all 'matter' is composed of fermions, while all 'interactions' are mediated by bosons. But *dark matter* could be something else; we simply do not know. Also, gravitation is thought by many physicists to be mediated by *gravitons*, a spin 2 boson. However, we simply do not know if that is correct. So, this axiom may be changed some day.

The Hilbert space of composite quantum entities is given in terms of the Hilbert spaces of the constituent bosons and fermions by a non-intuitive

---

[3]This group is usually realized as $SU(2)$, but that tends to hide its physical origin.

mathematical construction. As noted in Axiom 2, states are associated with the total Hilbert space of a quantum system, or more specifically with its von Neumann algebra. However, the partial trace does define a map from the states of the system to the states of any subsystem, but as is well known it does not map pure states to pure states. So, Axiom 3 is the place in this approach where partial trace should be introduced. Even though the partial trace is justifiably considered to be a non-commutative version of conditional expectation, it is not related to quantum conditional probability (see Section 2.6) or to quantum integrals (see Section 2.13). These remarks become quite relevant when considering entanglement in Chapter 3.

This axiom is included because of its importance in quantum theory. However, it is not presented in total detail since it is not going to play a role in this treatise, which focuses on probability and leaves spin, statistics and partial trace to a side. This axiom is a complicated condition that seems to have nothing to do with the other axioms, and it introduces spin into quantum theory in a seemingly *ad hoc* manner. If you like to think in terms of analogies, then the Fifth Axiom of Euclid and the Axiom of Choice come to mind. While generations have fretted over deeper explanations of the origin of probability in quantum theory, there is not much concern with an underlying explanation of spin that goes deeper than noting that spin comes from representation theory. However, left unanswered is why some irreducible representations correspond to the observed fundamental particles, while most do not. The reader can refer to the literature on spin, most notably in quantum field theory.

Another mathematical structure introduced in this axiom is that of a group representation. This is an important application of mathematics in quantum physics (see [45]). Nonetheless, it will not be playing any role in this treatise. The basics as presented here are not explicitly invariant under either representations of the Galilean group or of the Poincaré group, but rather consistent with either of these groups. Note that the idea that quantum theory automatically carries a representation of the Poincaré group led to the mistaken conviction that parity is conserved in all interactions in quantum theory. We also will not be considering quantum interactions explicitly, except in so far as to say here that they could be viewed as a type of quantum event or as certain terms in a Hamiltonian. Physical measurements will also be represented as a certain type of quantum event associated with the projection valued measure of a self-adjoint operator. See Section 2.3 for more details.

It may seem strange that the non-intuitive concept of spin enters an axiom, while the intuitive concept of position does not. Note that position, together with linear momentum, enter quantum theory via a representation of the Weyl–Heisenberg group. This could be taken as a shortcoming of the approach taken in this treatise. However, none of these particular observables will play a role here, though all of them could be present in a more complete axiomatization of quantum theory.

1.5   Axiom 4: Time Independent Born's Rule

Since a self-adjoint operator T is not a real number, it would seem that the notation $T \in B$, where $B \subset \mathbb{R}$ is a Borel set, should be non-sensical. However, we define $T \in B := P_T(B)$, a quantum event which is interpreted as the physical event that the observable T takes a value in the set B. This notation is used in the next axiom.

Let T be any self-adjoint operator (perhaps associated with an observable quantity of some physical entity), ρ be a density matrix and B be a Borel subset of $\mathbb{R}$. Then the *quantum probability* that T has a value in the set B given ρ is defined to be $P_\rho(T \in B) := l_\rho(P_T(B)) = Tr(P_T(B)\rho)$, where P_T is the pvm of the self-adjoint operator T and Tr is the trace of a trace class operator. ∎

While this is the form of Born's Rule most likely known to the reader, it will turn out, perhaps surprisingly, to be beside the main point of this treatise, which is that the important probabilities are those of multiple events. Here, we are considering only the probability of the single event $T \in B$.

This axiom, or a simple consequence of it about expected values, is part of standard quantum theory as found in the textbooks and as practiced in the scientific community. Only later on will we put time dependence into this and, most importantly, elevate the resulting formula to be the basic time evolution equation of quantum theory.

For fixed T and ρ the assignment $B \mapsto P_\rho(T \in B) \in [0,1]$ for B a Borel subset of $\mathbb{R}$ is a *probability measure* on $\mathbb{R}$ in the sense of measure theory, as we now prove.

To show the two normalizations of a probability measure, we note first that $P_\rho(T \in \emptyset) = Tr(P_T(\emptyset)\rho) = Tr(0\,\rho) = Tr(0) = 0$ and second that $P_\rho(T \in \mathbb{R}) = Tr(P_T(\mathbb{R})\rho) = Tr(I\,\rho) = Tr(\rho) = 1$.

To show σ-additivity let $\{B_j \mid j \in \mathbb{N}\}$ be a countable family of disjoint Borel subsets of $\mathbb{R}$. We can calculate the trace of a trace class operator using any orthonormal basis of $\mathcal{H}$. We choose an orthonormal basis $\{\phi_k\}$ which diagonalizes the trace class operator ρ. Such a basis, which is not unique, is guaranteed to exist by the spectral theorem. Specifically, $\rho\,\phi_k = \lambda_k\,\phi_k$ with all $\lambda_k \geq 0$ and $\sum_k \lambda_k = 1$. Then we have that

$$P_\rho\left(T \in \bigcup_j B_j\right)$$

$$= Tr\left(P_T\left(\bigcup_j B_j\right)\rho\right) = \sum_k \left\langle \phi_k, P_T\left(\bigcup_j B_j\right)\rho\,\phi_k\right\rangle$$

$$= \sum_k \left\langle \phi_k, \sum_j P_T(B_j)\,\rho\,\phi_k\right\rangle \quad \text{using strong operator topology}$$

$$= \sum_k \left\langle \phi_k, \sum_j P_T(B_j)(\lambda_k\,\phi_k)\right\rangle = \sum_k \sum_j \lambda_k\langle \phi_k, P_T(B_j)\,\phi_k\rangle$$

$$= \sum_j \sum_k \lambda_k\langle \phi_k, P_T(B_j)\,\phi_k\rangle \quad \text{using Fubini's theorem}$$

$$= \sum_j \sum_k \langle \phi_k, P_T(B_j)(\lambda_k\,\phi_k)\rangle = \sum_j \sum_k \langle \phi_k, P_T(B_j)\,\rho\phi_k\rangle$$

$$= \sum_j Tr\big(P_T(B_j)\,\rho\big)$$

$$= \sum_j P_\rho(T \in B_j).$$

Fubini's theorem applies since the terms in the double sum are non-negative.

We discuss mainly the case of pure states represented by $\phi \in \mathcal{H}$ where $\|\phi\| = 1$, in which case $\rho = |\phi\rangle\langle\phi|$. We then introduce the notation

$$P_\phi(T \in B) := Tr(P_T(B)\,|\phi\rangle\langle\phi|) = \langle\phi, P_T(B)\phi\rangle,$$

where the left side is read as: The probability given the pure state represented by ϕ that T has a value in B. There are alternative formulas for Born's Rule for the case of pure states. Here are two useful ones:

$$P_\phi(T \in B) = \|P_T(B)\phi\|^2 = Tr(P_T(B)E_\phi).$$

Here, $E_\phi := |\phi\rangle\langle\phi|$ in Dirac notation is a rank 1 projection. Note that E_ϕ is also a quantum event if the von Neumann algebra is a Type I factor. Of course, it is always a density matrix.

Axiom 4 is the simplest form of Born's Rule. It will be generalized and as such its central importance in quantum theory will become apparent. Note that *quantum probability* in this treatise simply means all possible forms of Born's Rule, which is all the formulas for calculating quantum probabilities. In other contexts quantum probability has more general formulas. But in all cases what I mean by a *probability theory* is a mathematical formalism that assigns numbers in the unit interval $[0, 1]$ to temporally ordered sequences of events. As such the formalism must include a rigorous mathematical definition of what an event is. Note that this clearly distinguishes quantum probability, as presented here, from many theories dubbed *non-commutative probability*[4], which typically study the properties of a state defined on some non-commutative algebra, such as a C^*-algebra. While a state corresponds to an integral with respect to a probability measure in the commutative case, those theories usually do not involve events nor their probabilities.

The *physical significance* of quantum probability is that it corresponds to the *relative frequency* of *physical events* and, in particular, to the empirically observed quantities associated to a self-adjoint operator and a state. Even though this statement might be deemed to be philosophical, it is consistent with the practice of physics as a *quantitative* discipline, both experimentally and theoretically. Quantum probability does not give qualitative results, but rather numbers, which are more than tendencies, propensities, inclinations or criteria for decision making. They are quantitative data which can be checked by experiment. The point is that in experiments one measures *numerical quantities* which serve as the gold standard for comparison with theoretical numerical computations.

Let me emphasize again that quantum probability applies to *all* events, not just those observed directly in experiment.

1.6   Axiom 5: Dynamics

Every physical system has two associated actions of one-parameter groups. These are denoted as S_t and E_t, where $t \in \mathbb{R}$ is considered as a parameter in the theory which corresponds to time. (By *one-parameter group* we mean

---

[4]Quantum probability is defined using the events in a non-commutative von Neumann algebra and thus merits being called a non-commutative probability theory.

that $S_a S_b = S_{a+b}$ for all $a, b \in \mathbb{R}$ and that $S_0 = id$, the appropriate identity map. Similar formulas hold for E_t.)

The one-parameter group S_t maps the convex set $\mathcal{S}$ of *all* states (including mixed states) to itself, while the one-parameter group E_t maps the set $\mathcal{E}$ of all quantum events to itself.

The *dynamics (or time evolution)* of the physical system itself for an initial observable represented by a self-adjoint operator $T = T^*$ and an initial density matrix ρ is given by the *time dependent Born's Rule*

$$t \mapsto Tr(E_t(P_T(B))\, S_t \rho) \quad \text{for } t \in \mathbb{R}. \tag{1.6.1}$$

We assume that *conservation of probability* holds, which by definition means that for fixed ρ and fixed T the mapping $B \mapsto Tr(E_t(P_T(B))\, S_t \rho)$ is a probability measure for every $t \in \mathbb{R}$. Here, *initial* means at time $t = 0$. ∎

The reason for having two one-parameter groups in (1.6.1) is in order to accommodate various 'pictures' (Schrödinger, Heisenberg, interaction) in the axiomatics. This will be exemplified in the next section.

Following tradition I have stated these as five separate axioms. Nonetheless, logically speaking Axioms 1 and 2 form one statement on kinematics, while Axioms 4 and 5 are one statement on dynamics. Axiom 3 as noted above stands out as being quiet different, but in a technical sense it is also a statement on kinematics — although a long, convoluted statement.

A special case of Axiom 5 is when S_t maps the set of pure states to itself. Then we have that

$$S_t |\phi\rangle\langle\phi| = |\phi_t\rangle\langle\phi_t| \tag{1.6.2}$$

for some unit vector $\phi_t \in \mathcal{H}$ which is not uniquely determined, but is unique modulo a phase factor. In this case the dynamics is given by another version of Born's Rule, which is a special case of (1.6.1):

$$t \mapsto Tr(E_t(P_T(B))\, |\phi_t\rangle\langle\phi_t|) = \langle \phi_t, E_t(P_T(B))\phi_t \rangle = \|E_t(P_T(B))\phi_t\|^2, \tag{1.6.3}$$

where we again assume that conservation of probability holds.

1.7 Models of the Axioms

The time evolution of neither the events (given by E_t) nor of the states (given by S_t) is fundamental. What is fundamental in the sense that it corresponds to observations is the combination of these two one-parameter groups in the above expressions for the time dependent probability given by Born's Rule. Neither of these two one-parameter groups is uniquely determined by the axioms. There are various *models* of these axioms for which S_t and E_t are quite different. Some of these models are typically called *pictures* in the literature. Axiom 5 was written to include the three most commonly used models, which we present next. But other models are possible.

I am not the first to say that quantum theory is a way for calculating probabilities, and nothing else. But I am noting that different models have different ways for doing those calculations. Moreover, the fundamental, final formula for arriving at those calculations is the same in all models (namely, the Generalized Born's Rule), which is *covariance*, and that the resulting number is the same in all equivalent models, which is *invariance*.

The most commonly used model is called the *Schrödinger picture*, but I prefer to call it the *Schrödinger model*. This model has the property that $E_t = id_{\mathcal{E}}$ for all $t \in \mathbb{R}$, that is, the events have trivial time evolution. One also says that the observables, which are events since they are pvm's, are *time independent* in this model. In this model the time evolution maps pure states to pure states and is given by *Schrödinger's equation*. The solution of this equation with an initial condition ϕ at time $t = 0$ is given by functional analysis as $\phi_t = e^{-itH} \phi$, where $H = H^*$, a possibly unbounded operator acting in $\mathcal{H}$, is the Hamiltonian in Schrödinger's equation. (The minus sign in the exponent of e^{-itH} is purely conventional. It has no physical significance; it does not need to be 'explained'. Also, recall that $\hbar = 1$.) Taking the initial condition ϕ to be a unit vector, it follows by the unitarity of e^{-itH} that ϕ_t is a unit vector for all $t \in \mathbb{R}$. Then S_t in the Schrödinger model is defined on pure states by the formula (1.6.2), and the time dependent Born's Rule (1.6.3) gives the probability that the observable $T = T^*$ is in the Borel subset B of $\mathbb{R}$ as a function of time $t \in \mathbb{R}$ as

$$t \mapsto \langle \phi_t, P_T(B)\phi_t \rangle. \tag{1.7.1}$$

Since ϕ_t is a unit vector, for each fixed time t this is a probability measure as we have seen earlier. That is to say, conservation of probability holds in the Schrödinger model.

Note that Schrödinger's equation is pushed into the background even in the Schrödinger model. All that is important is the Hamiltonian H which in and of itself specifies the flow $t \mapsto \phi_t = e^{-itH} \phi$ in the Hilbert space $\mathcal{H}$. However, there is a preference for thinking that differential equations are basic and that their solutions, in this case the flow, are secondary. Of course, to understand this flow in specific cases it often is a good idea to solve, or at least analyze, Schrödinger's equation. In introductory texts it is usually considered to be pedagogically advantageous to give Schrödinger's equation a central role. And I do this in [39] for example. But this should be taken by those with more knowledge with a ton, not a grain, of salt, since (1.7.1) is the fundamental time evolution equation for one observable in the Schrödinger model for pure states. For those who prefer differential equations, the time derivative of (1.7.1) formally gives for any Borel subset B of $\mathbb{R}$ that

$$\frac{d}{dt}\langle\psi, P_T(B)\psi\rangle = \langle\psi, i[H, P_T(B)]\,\psi\rangle. \tag{1.7.2}$$

where $\psi = \phi_t$ and $[R, S] := RS - SR$ is the *commutator* of the two operators R, S. This is an easy non-rigorous exercise, provided domain considerations are ignored. As is typical, the differential equation (1.7.2), which in no way do I wish to consider to be fundamental, is a more singular object than its 'solution' (1.7.1), which I do consider to be fundamental. The mapping $B \mapsto i[H, P_T(B)]$ maps the empty set $\emptyset$ to 0 and is formally σ-additive . But it also maps $\mathbb{R}$ to 0, and so it is not a monotone self-adjoint operator valued measure, even if H is a bounded operator. On the other hand, (1.7.2) is formally valid in the Heisenberg and interaction models, as described below. It seems that for Born's Rule there is no 'nice' differential equation that holds (even formally) in all models.

Next, we extend S_t to act on density matrices ρ as follows. First, by the spectral theorem there exists an orthonormal basis $\{\phi_k\}$ of $\mathcal{H}$ and real numbers $\lambda_k \geq 0$ such that $\sum_k \lambda_k = 1$ and $\rho = \sum_k \lambda_k |\phi_k\rangle\langle\phi_k|$. We also define $U_t := e^{-itH}$, the unitary time evolution operator, for all $t \in \mathbb{R}$. The time evolution of the ket is $|\phi_k\rangle \mapsto U_t|\phi_k\rangle$, while that of its dual bra is $\langle\phi_k| \mapsto \langle\phi_k|U_t^*$. So, $|\phi_k\rangle\langle\phi_k| \mapsto U_t|\phi_k\rangle\langle\phi_k|U_t^*$. Putting this together, the time evolution of ρ is defined by

$$\rho \mapsto \sum_k \lambda_k\left(U_t|\phi_k\rangle\langle\phi_k|U_t^*\right) = U_t\left(\sum_k |\lambda_k\phi_k\rangle\langle\phi_k|\right)U_t^* = U_t\,\rho\,U_t^*.$$

The Schrödinger model is so widely used that properties specific to it are often thought to be general properties of quantum theory. I will discuss this misunderstanding in more detail later on.

Another model is called the *Heisenberg picture*, which I prefer to call the *Heisenberg model*. In this model one has $S_t = id_\mathcal{S}$ for all $t \in \mathbb{R}$. One says that the states are *time independent* in this model. In particular, S_t maps pure states to pure states. The time evolution of a quantum event $P \in \mathcal{E}$ is given in this model by the *Heisenberg equation*

$$P \mapsto E_t\, P := U_t^* P U_t, \qquad\qquad (1.7.3)$$

where U_t for $t \in \mathbb{R}$ is a strongly continuous unitary group acting on $\mathcal{H}$, provided that $U_t^* P U_t \in \mathcal{E}$ for all $t \in \mathbb{R}$. It is important to note that this flow is basic; it is not arrived at as the solution of a differential equation. Next it follows by Stone's theorem that $U_t := e^{-itH}$ with $H = H^*$, a possibly unbounded self-adjoint operator. (The minus sign in the exponent of e^{-itH} is again purely conventional. It has no physical significance; it does not need to be 'explained'.) At this point a technical detail arises, since we are not necessarily dealing with a Type I factor. In order to consider that H corresponds to a physical observable, it is reasonable to impose the condition (as was done in Axiom 1) that also all the events $P_H(B)$, where $B \subset \mathbb{R}$ is Borel, correspond to physical observables and therefore are in the von Neumann algebra $\mathcal{V}$. This then implies that $f(H) = \int_\mathbb{R} f(\lambda)\, dP_H(\lambda) \in \mathcal{V}$ for all bounded, Borel functions $f : \mathbb{R} \to \mathbb{C}$. In particular, $U_t = e^{-itH} \in \mathcal{V}$ for all $t \in \mathbb{R}$. Therefore the map in (1.7.3) sends $\mathcal{E}$ to $\mathcal{E}$.

   If P_A is the pvm of a self-adjoint operator A, then the same time evolution applies to it:

$$P_A \mapsto E_t\, P_A := U_t^* P_A U_t,$$

where $U_t^* P_A U_t(B) := U_t^* P_A(B) U_t$ for all Borel subsets B of $\mathbb{R}$. It turns out from functional analysis that this time evolution maps a pvm, which is a family of projections, to another family of projections, which turns out itself to be a pvm. It follows that the pvm P_A of a self-adjoint operator maps to the pvm of another self-adjoint operator. Explicitly, one can show that

$$E_t\, P_A = U_t^* P_A U_t = P_{U_t^* A U_t}.$$

   Again, the Heisenberg model has certain specific properties which are not general properties of quantum theory. Such a particular property of the Heisenberg model is that E_t extends naturally to a time evolution of $\mathcal{L}(\mathcal{H})$ defined by $E_t\, T := U_t^* T U_t$ for all $T \in \mathcal{L}(\mathcal{H})$. The group E_t in any of these manifestations is difficult to accommodate with ordinary intuition, since it says that observables are time dependent, which might seem sensible enough if one is speaking of position or angular momentum. However, events are self-adjoint operators; so they are observables too. So, in the Heisenberg model events are time dependent! On the other hand states, which intuitively tell us everything about a system at any moment, are time independent. This is backwards from the common intuition of what 'events' are and what 'states' are. In part this is due to a poor choice in terminology. A quantum event $E \notin \{0, I\}$ is an observable that has spectrum $\{0, 1\}$. This corresponds to two possible observed values. Sometimes an event is called a *Yes-No experiment*. We tend to think that the value 1 (Yes) means that the event occurred, while the value 0 (No) means that the event did not occur. But this is misleading; an event is an observable that can give either of these two values. In either case the event has been observed. In any event (in another sense yet of that word) I find quantum theory to be non-intuitive.

Also notice that the time evolution of events in the Schrödinger model extends naturally to $\mathcal{L}(\mathcal{H})$ by setting $E_t := I_{\mathcal{L}(\mathcal{H})}$ for all $t \in \mathbb{R}$. This particular extension property is not a necessary aspect of quantum theory.

To show that conservation of probability holds in the Heisenberg model let ρ be a density matrix and $T = T^*$ be a self-adjoint operator. Then

$$Tr(E_t(P_T(B))\, \rho) = Tr(U_t^* P_T(B) U_t \rho) = Tr(P_T(B)\, U_t \rho U_t^*)$$

$$= Tr(P_T(B)\rho_t), \qquad (1.7.4)$$

where $\rho_t := U_t \rho U_t^*$, the (Schrödinger model time evolved) density matrix for every $t \in \mathbb{R}$, is also a density matrix. But we have already proved in our discussion of the Schrödinger model that the last expression in (1.7.4) gives a probability measure on $\mathbb{R}$.

The same Born's Rule (1.6.3) is used in the Schrödinger model and the Heisenberg model. What that means is that the same *formula* is used in both. This is *covariance*. But more is true. The two models give the same numerical probability using (1.6.3) provided that the same self-adjoint operator H is used in both models. This is *invariance*. (In other words, *compatible* sign conventions are being used in the two models. This does have a significance, since it makes the following proof work.)

Here is the proof of this well-known, yet important result. We assume that in both models at time $t = 0$ the state is represented by ϕ and that the pvm is P_T. Of course, in the Schrödinger model only the state can be time dependent, while in the Heisenberg model only the pvm can be time dependent. Be aware that all of the following expressions are time dependent. Then for all $t \in \mathbb{R}$ we have in an obvious notation that

$$P_{\phi_t}^{Sch}(T \in B) = ||P_T(B)\,\phi_t||^2 = ||P_T(B)\,e^{-itH}\,\phi||^2$$
$$= ||e^{itH}P_T(B)\,e^{-itH}\,\phi||^2 = ||U_t^* P_T(B) U_t\,\phi||^2$$
$$= ||E_t P_T(B)\phi||^2 = P_\phi^{Heis}(E_t T \in B).$$

This is, of course, a special case of the following calculation. Let ρ be a density matrix. Then we see that

$$P_{\rho_t}^{Sch}(T \in B) = Tr(P_T(B)\,\rho_t) = Tr(P_T(B)\,U_t\rho U_t^*) = Tr(U_t^* P_T(B)\,U_t\rho)$$
$$= Tr(E_t P_T(B)\,\rho) = P_\rho^{Heis}(E_t T \in B). \tag{1.7.5}$$

This calculation should be compared with (1.7.4).

But still other models are used for which both S_t and E_t are non-trivial. The *interaction pictures* are such models. There are a multitude of such models. One starts by taking any one-parameter *family* of unitary operators U_t for $t \in \mathbb{R}$ acting on the Hilbert space $\mathcal{H}$. Next one changes the Schrödinger model to become the new *interaction model* by transforming the pure states by $\psi \mapsto U_t\psi =: \psi'$ and mapping the pvm's by $P_S \mapsto U_t P_S U_t^* =: P_{S'}$, where S is a self-adjoint operator and $U_t S U_t^* =: S'$. One then sees that

$$P_\psi(S \in B) = \langle \psi, P_S(B)\psi \rangle = \langle U_t\psi, U_t P_S(B)U_t^* U_t\psi \rangle = P_{\psi'}(S' \in B).$$

This shows that the time dependent probability as calculated in each model is exactly the same even though the time dependence of both the states and the pvm's has been changed. The transformation from the Schrödinger model to the Heisenberg model is a special case of this. Another very special case is to take $U_t := U$ for all $t \in \mathbb{R}$ where U is a fixed unitary transformation.

Note that in the interaction model U_t need not be a unitary group nor does $t \mapsto U_t$ need to have any sort of continuity. The lack of continuity, for example, may seem to you to be a mathematical trick with no physical intuition behind it. If so, you are right. The interaction model is just used as a convenient mathematical technique to help one calculate probabilities, and its intermediary steps have no physical significance.

All of these models are *isomorphic* (defined below) to the Schrödinger model, but there are other models which are not. The point of the axioms is to capture certain features, which are to be considered as basic to quantum theory. They are not meant to be *categorical*, in the same sense as, for example, the axioms of Euclidean plane geometry are meant to describe completely its limited topic. Rather the axioms for quantum theory are meant to be like the axioms in mathematics of a group, of which there are many non-isomorphic objects. Similarly, as we shall discuss in detail later, there are many non-isomorphic quantum theories. Here is that important definition, where $\text{Bij}(A)$ denotes the set of bijections of a set A to itself.

Definition 1.7.1. A *model of quantum theory* is a quadruple $(\mathcal{H}, \mathcal{V}, E, S)$ where $\mathcal{H}$ is a Hilbert space such that $\mathcal{V} \subset \mathcal{L}(\mathcal{H})$ is a von Neumann algebra, $E : \mathbb{R} \to \text{Bij}(\mathcal{E})$ is a one-parameter group of bijections of the set $\mathcal{E}$ of events in $\mathcal{V}$ to itself, and $S : \mathbb{R} \to \text{Bij}(\mathcal{S})$ is a one-parameter group of bijections of the set $\mathcal{S}$ of states in $\mathcal{T}(\mathcal{H})$ to itself such that for every $t \in \mathbb{R}$ the *time dependent Born's Rule*

$$t \mapsto Tr(E_t(P_T(B))\, S_t\rho)$$

is a probability measure on $\mathbb{R}$ for every density matrix ρ, for all self-adjoint operators T and all Borel subsets B of $\mathbb{R}$. As noted earlier this last condition is called *conservation of probability*.

An *isomorphism* of the quantum theories $(\mathcal{H}, \mathcal{V}, E, S)$ and $(\mathcal{H}', \mathcal{V}', E', S')$ is a *one-parameter family* of onto unitary transformations $U_t : \mathcal{H} \to \mathcal{H}'$ with $U_t(\mathcal{V}) = \mathcal{V}'$ which satisfy condition (1.7.6) below and the parameter is $t \in \mathbb{R}$. In the model $(\mathcal{H}, \mathcal{V}, E, S)$ we let $T = T^*$ be affiliated to $\mathcal{V}$ and ρ be a density matrix. These correspond in the usual way in the model $(\mathcal{H}', E', \mathcal{V}', S')$ to the self-adjoint operator $T' := U_t T U_t^*$ affiliated to $\mathcal{V}'$ and to the density matrix $\rho' := U_t \rho U_t^*$. Then for all $t \in \mathbb{R}$ and for all Borel subsets B of $\mathbb{R}$ we require

$$Tr(E_t(P_T(B))\, S_t\rho) = Tr(E_t'(P_{T'}(B))\, S_t'\rho'), \qquad (1.7.6)$$

which is called *preservation of probability*. (This condition is based on the equality of probabilities (1.7.5) for the Schrödinger and Heisenberg models.)

Finally, we say that two models of quantum theory are *isomorphic* if there exists an isomorphism between them.

Please be careful to note the difference between *conservation of probability* and *preservation of probability*. The definition of isomorphism of models must be modified in an obvious way to include preservation of probability for consecutive and conditional probabilities after these are defined later on.

A special case of isomorphism of models is when U_t is a unitary transform that does not depend on 'time' t. A simple example is the Fourier transform $\mathcal{F} : L^2(\mathbb{R}^3) \rightarrow L^2(\mathbb{R}^3)$, which transforms the spatial representation into the momentum representation. While ψ in the domain of $\mathcal{F}$ tells us, to use common parlance, 'where the electron is' (at least in a probabilistic sense), its thoroughly equivalent representation $\mathcal{F}\psi$ tells us 'how the electron is moving' (again at least in a probabilistic sense). The moral of this little story is that the 'wave function' of an electron does not have either of these quoted expressions as its model independent meaning, although each of them does have a model dependent meaning.

There will surely be those who wish to maintain that the time changing state and the Schrödinger model are correct while the corresponding theory in the Heisenberg model is not correct. This just amounts to rejecting the importance of the previous definition. In that case the test is to design an experiment whose results would be different in the two models. And then do the experiment to see which model is wrong. (Maybe both will be wrong!) But the Schrödinger model as presented here would have to be augmented with another measurable property instead of the Generalized Born's Rule, which is the same in the two models. In other words one would have to propose a property that only holds in the Schrödinger model.

1.8   How Many Time Evolutions?

A common criticism of standard quantum theory is that it has two time evolutions: one from Schrödinger's equation and the other from the collapse of the state. It is widely held that neither of these can be the consequence of the other. After all, Schrödinger's equation gives a continuous, deterministic time evolving state while the collapse is discontinuous and probabilistic.

Of course, one could try to explain collapse as a continuous change that is so quick as to appear discontinuous. Or, on the other hand, one might try to explain continuous time evolution as a rapid succession of discontinuous jumps, much as a motion picture is not a picture of continuous motion but rather many static pictures in succession giving the impression of motion. While these remain as logical possibilities, neither is readily implementable.

So, leaving these possibilities to a side, these two time evolutions are not only incompatible, but also neither can be *the* basic description of time evolution in quantum theory. There are logically then two alternatives. The first is to accept that nature is described by two independent time evolutions, and that's the end of the matter. The second is that there is one time evolution which is somehow more basic than these two. I am advocating for the second option in a rather specific way.

1.9 Relation to Observation

It is crucial to understand in this formalism the relation of quantum theory to physical observations is based exclusively on those probability statements in the theory that are model independent within an isomorphism class of models. Unitary transformations are the isomorphisms of Hilbert spaces, and the isomorphisms they induce on von Neumann algebras are known as *spatial isomorphisms*. Physically, this means that the formulation of a theory in the context of one specific von Neumann algebra can always be translated into the context of a different spatially isomorphic von Neumann algebra via the application of a unitary transformation of Hilbert spaces. While such a unitary transformation changes the model being used to understand the physics, only properties that hold in all isomorphic models can possibly be relevant to understanding physical phenomena. A property that holds in one model, but not in all other isomorphic models, must absolutely never be considered as having any physical significance. Such properties can be useful as intermediate steps in the analysis of a physical system, but nothing more than that. Such model dependent properties do not need to be 'interpreted' nor 'explained' *vis a vis* their physical significance. They simply have no physical significance.

As an example, a Schrödinger operator acting in $L^2(\mathbb{R}^3)$ could have an eigenstate which is represented by a C^∞ function. That function corresponds to a unit vector in any isomorphic Hilbert space, whose elements need not be represented by functions on a differential manifold. So, the concept of C^∞ does not apply in that isomorphic model. Nonetheless, the eigenvalue of the corresponding self-adjoint operator for the corresponding eigenstate is the *same* real number for all isomorphic models and has a physical significance, provided that the original Schrödinger operator represents a physical system. For example, the *Segal–Bargmann space* on $\mathbb{C}^3$ (see [2]) can be used instead of $L^2(\mathbb{R}^3)$. (A unitary transformation effecting this change of model is the *Segal–Bargmann transform*.) In the Segal–Bargmann model all pure states are represented by C^∞ functions. So, in that model the property of a pure state having a C^∞ representative is unremarkable. The failure to recognize this elementary aspect of quantum theory leads to pointless discussions about the 'meaning' or 'interpretation' of model dependent properties.

1.10   An Incomplete Theory

Many important properties of quantum theory have been omitted from the axioms. This is not because they lack importance. Rather I think they are not needed for now, but they can and obviously should be part of a more complete theory. These include, but are not limited to, complementarity and uncertainty. However, particle properties, having probability one and persisting in time, remain clearly outside of the scope of this treatise, which treats transient events with probabilities in $[0, 1]$. Thus, this indicates a need to extend this theory, perhaps to quantum fields. Quantum phenomena, such as continuous measurements or stochastic processes, also remain for future consideration. This approach to quantum theory is not complete.

Chapter 2

Quantum Probability

le hasard . . .
qui est en réalité
l'ordre de la nature

— Anatole France

2.1 Introduction

This lengthy chapter is meant to motivate, define and study the basics of quantum probability. This will motivate the axioms in Chapter 1 and lead to a modification of them. I start by explaining how certain classical probability measures (in the sense of the formulation of Kolmogorov in his book [25]) arise as a consequence of various hypotheses, which are accepted generally by physicists who work in quantum theory. The idea that probability is an added-on 'interpretation' of quantum theory is a misconception, that leads many to think that quantum probability can be replaced by using some alternative 'interpretation'. While this has been known since at least the publication of von Neumann's book [41] in 1932, it seems not to be well known, neither in the mathematics nor the physics communities. Actually, just about everything in this chapter can be found in the literature, with the unique exception of the Generalized Born's Rule as the fundamental time evolution equation of quantum theory. I have tried to re-organize this known, but not well known, material which has too often been overlooked by people who were, I am afraid, more interested in 'interpreting' quantum theory than in understanding it.

Before going into details it is necessary to understand what is meant by a *probability theory* in the most general terms. But this is rather clear. It is any mathematical theory with objects (*theoretical events*) together with ways of assigning real numbers lying in the closed interval $[0, 1]$ to single events and, most importantly, to combinations of events. The application of probability theory to experimental science is that these real numbers, called *probabilities*, have the significance that they describe *relative frequencies* of certain *physical events*. There are many more details on both the theoretical and experimental sides, but this is the basic idea. The approach, known as *Classical Probability*, given by Kolmogorov in [25] is in this view just one of many possible probability theories. We will later see its relation to Quantum Probability.

I understand *Quantum Probability* to mean that part of *Quantum Theory* which gives rise to probabilities and the rules that apply to them. As such it is not an independent theory, since it is essentially linked to the rest of Quantum Theory. In other words, Quantum Probability is a part of physics. It also is an example, (the first historically, I guess), of a Non-commutative Probability Theory, which is an ongoing research area in basic mathematics. But I shall not deal with this more general mathematical topic, except to note that it (typically?) does not come as part and parcel of a theory with time evolution nor is it intended to be a physical theory. For a glimpse into some beautiful advanced topics in Quantum Probability beyond the Generalized Born's Rule see [1], [34], [38] and references therein.

2.2   The Physical Assumptions

A given quantum system is described by a formulation based on a specific complex, non-zero Hilbert space $\mathcal{H}$ which is often, but not necessarily, infinite dimensional. This is implicit in Dirac's notation, though it is usually accepted explicitly by those in the quantum physics community. Note that the case when $\mathcal{H}$ has dimension 1 is included for now, even though the resulting quantum physics is trivial. But in this chapter we do exclude the case $\mathcal{H} = 0$, since that leads to no physics whatsoever, since there are no states and only one event. Actually by Axiom 1, $\dim \mathcal{H} \geq 2$ in most of this treatise.

Within such a mathematical model quantum physics is assumed by most in the physics community to satisfy these properties among others:

- The state of a quantum system is described by vectors $\psi \in \mathcal{H}$ with $||\psi|| = 1$ with two such unit vectors ψ_1, ψ_2 describing the same state provided that there exists $\lambda \in \mathbb{C}$ such that $\psi_1 = \lambda \psi_2$. (Necessarily $|\lambda| = 1$.)

- Most physical measurements (with the remarkable exception of time measurements) are represented by self-adjoint, densely defined linear operators acting on a linear subspace of $\mathcal{H}$.
- The same physical measurement performed on an ensemble of quantum systems, which are all described by the same state, does not in general yield the same measured value, but rather values with some relative frequencies of numbers in subsets of $\mathbb{R}$.

Various comments are in order. The first property only describes *pure states* and not the more general *mixed states*, which also enter into quantum theory. However, what we will be saying is easily generalized to the setting with all states, including the mixed states, under consideration.

The second property may have a converse, namely that every self-adjoint, densely defined linear operator corresponds to a physical measurement. This converse is denied by theories with *superselection rules*. But this possibility does not concern us, since we will not be using the converse statement.

The third property concerns what one observes in experiment. The only way to deny it is to re-define the meaning of the word 'same'. But we will not use this property is the following argument, but rather offer a explanation of it using classical probability theory.

2.3   The Mathematical Model

The central mathematical fact is that there is a complete description of densely defined, self-adjoint operators which act in a Hilbert space. This is the content of the *Spectral Theorem*. This description is given in terms of *projection-valued measures*, which we will define momentarily. However, this section is not a complete presentation of the appropriate functional analysis, which I expect the reader already to know. Mainly it serves to establish notation and as a quick review. See [42] for full details.

First, we recall some basics from Hilbert space theory. We let $\langle \cdot, \cdot \rangle$ denote the inner product in $\mathcal{H}$, a complex Hilbert space, and $|| \cdot ||$ its associated norm. One has $||\psi|| := \langle \psi, \psi \rangle^{1/2}$ for all $\psi \in \mathcal{H}$. A function $T : \mathcal{H} \to \mathcal{H}$ is said to be *linear* if, just as in linear algebra, we have

$$T(\alpha_1 \psi_1 + \alpha_2 \psi_2) = \alpha_1 T(\psi_1) + \alpha_2 T(\psi_2)$$

for all $\psi_1, \psi_2 \in \mathcal{H}$ and all $\alpha_1, \alpha_2 \in \mathbb{C}$. Furthermore, we say that such a function is a *(linear) operator*. If such an operator T has the property that there exists some real number $M \geq 0$ such that $||T\psi|| \leq M||\psi||$ holds for all $\psi \in \mathcal{H}$, then we say that the operator T is *bounded*.

A *projection* $E : \mathcal{H} \to \mathcal{H}$ is by definition a linear operator such that E is idempotent (meaning that $E^2 = E$) and E is self-adjoint (meaning that $E^* = E$). Projections are bounded operators. It is important to note that 0, the zero operator (i.e., the operator which maps every $\psi \in \mathcal{H}$ to $0 \in \mathcal{H}$) is a projection. Also, $I_\mathcal{H}$, the identity operator acting on $\mathcal{H}$ is a projection, where by definition $I_\mathcal{H}\psi := \psi$ for all $\psi \in \mathcal{H}$. Also, we define

$$\mathcal{L}(\mathcal{H}) := \{T : \mathcal{H} \to \mathcal{H} \mid T \text{ is linear and bounded}\}.$$

For each $T \in \mathcal{L}(\mathcal{H})$ we define its *operator norm* as

$$||T||_{op} := \sup \{||T\psi|| \mid \psi \in \mathcal{H} \text{ and } ||\psi|| \leq 1\}.$$

With this norm $\mathcal{L}(\mathcal{H})$ becomes a complete, normed space, which is to say that it is a *Banach space*. (Complete means that every Cauchy sequence with respect to the metric associated to the norm is convergent to an element in $\mathcal{L}(\mathcal{H})$. The *associated metric* is defined by $d_{op}(T_1, T_2) := ||T_1 - T_2||_{op}$ for all $T_1, T_2 \in \mathcal{L}(\mathcal{H})$.)

As examples we mention that for any projection $E \neq 0$ we have that $||E||_{op} = 1$. Also, $||I_\mathcal{H}||_{op} = 1$ since $\mathcal{H} \neq 0$.

Recall from standard measure theory that the smallest σ-algebra, which is denoted as $\mathcal{B}(\mathbb{R})$, of subsets of $\mathbb{R}$ that contains all finite open intervals (a, b) (with $a, b \in \mathbb{R}$ and $a < b$) is called the *Borel σ-algebra* of $\mathbb{R}$. Also, a set in $\mathcal{B}(\mathbb{R})$ is called a *Borel set*. Note that the empty set $\emptyset$ and the whole real line $\mathbb{R}$ are Borel sets.

We now have enough to give this definition[1].

Definition 2.3.1. A *projection-valued measure (pvm)* associated to a Hilbert space $\mathcal{H}$ is a function $P : \mathcal{B}(\mathbb{R}) \to \mathcal{L}(\mathcal{H})$ satisfying these properties:

- $P(B) \in \mathcal{L}(\mathcal{H})$ is a projection for every Borel set $B \in \mathcal{B}(\mathbb{R})$.
- $P(\emptyset) = 0$, the zero operator, and $P(\mathbb{R}) = I_\mathcal{H}$, the identity operator acting on $\mathcal{H}$.
- Let $\{B_j \mid j \in J\}$ be a finite or countably infinite family of Borel sets in $\mathcal{B}(\mathbb{R})$, which is disjoint (meaning that $B_j \cap B_k = \emptyset$ whenever $j \neq k$).

---

[1]Since projections are quantum events, I called this a *quantum event valued measure* or *qevm* in [39]. This was meant to make physics students feel more at home with a quite abstract mathematical structure.

Then we have the *σ-additivity condition*

$$P\left(\bigcup_{j\in J} B_j\right) = \sum_{j\in J} P(B_j),$$

in the sense that

$$P\left(\bigcup_{j\in J} B_j\right)\psi = \sum_{j\in J} P(B_j)\psi$$

holds for every $\psi \in \mathcal{H}$, where the possibly infinite sum on the right side is understood to mean the convergence with respect to the metric topology on $\mathcal{H}$ induced by its norm. That metric, denoted as d, is defined by $d(\psi_1, \psi_2) := \|\psi_1 - \psi_2\|$ for all $\psi_1, \psi_2 \in \mathcal{H}$. Also notice that $\bigcup_{j\in J} B_j$ is a Borel set, by definition of σ-algebra, and so the left side is also well defined.

A curious consequence of this definition is that for every pvm P we have $P(B \cap C) = P(B)P(C)$ for all $B, C \in \mathcal{B}(\mathbb{R})$. And this in turn implies that the family of operators $\{P(B) \mid B \in \mathcal{B}(\mathbb{R})\}$ is commutative.

The point of measure theory is to use measures in order to develop a well behaved theory of integrals. In courses this is typically done with measures with non-negative real values (such as *Lebesgue measure*), though also the generalization to measures with all real values or with complex values is sometimes presented. These generalizations are essentially the same as the theory with non-negative valued measures. Actually, this also works out for a measure $\mu : \mathcal{B}(\mathbb{R}) \to X$, where X is any Hausdorff topological vector space. The topology gives a meaning to convergent infinite sums in X. In particular, for any Borel measurable, function $f : \mathbb{R} \to \mathbb{C}$ we can consider whether the integral $\int_{\mathbb{R}} f(\lambda)\, d\mu(\lambda)$ exists as a uniquely defined element in X. The usual technical, measure-theoretic details apply. In the case of interest here the vector space is the complex vector space $\mathcal{L}(\mathcal{H})$, but not with its norm topology. Instead, we use the *strong operator topology* on $\mathcal{L}(\mathcal{H})$. Without going into all the technical details, let's simply note that a *sequence* of operators $A_n \in \mathcal{L}(\mathcal{H})$ converges to an operator $A \in \mathcal{L}(\mathcal{H})$ in the strong operator topology if and only if for every $\phi \in \mathcal{H}$ the sequence $A_n\phi \in \mathcal{H}$ converges to $A\phi \in \mathcal{H}$ in the norm topology of $\mathcal{H}$. The strong operator topology is the same as the operator norm topology on $\mathcal{L}(\mathcal{H})$ if and only if the dimension of $\mathcal{H}$ is finite. Using the norm topology in the infinite dimensional case gives an integral inadequate for *spectral theory* and hence inadequate for quantum theory.

A pvm P is a measure taking values in $\mathcal{L}(\mathcal{H})$, a Hausdorff topological vector space space when endowed with the strong operator topology. And so we have a well defined (and well behaved in the sense of measure theory) integral $\int_{\mathbb{R}} f(\lambda) \, dP(\lambda) \in \mathcal{L}(\mathcal{H})$ for any *essentially bounded*, Borel function $f : \mathbb{R} \to \mathbb{C}$. Also, an integral can even be defined for some unbounded Borel functions, but that leads to ever more technical details, which the reader can find in any good functional analysis text. Unfortunately, in the statement of the Spectral Theorem, which we state next, the integrand is an unbounded Borel function in the important case when the operator is not bounded.

Theorem 2.3.1 (Spectral Theorem). *Suppose that A is a densely defined self-adjoint operator acting in a Hilbert space $\mathcal{H}$. Then there exists a unique pvm $P_A : \mathcal{B}(\mathbb{R}) \to \mathcal{L}(\mathcal{H})$ such that*

$$A = \int_{\mathbb{R}} \lambda \, dP_A(\lambda). \tag{2.3.1}$$

A measurement of A with a value in a Borel set B (possibly a small interval) is represented by the quantum event $P_A(B)$, as is explained in Section 2.15.

Many other comments are in order, including why this theorem is so named. As a start we have the annoying fact that the function that we are integrating in (2.3.1) is the function $f(\lambda) = \lambda$ for all $\lambda \in \mathbb{R}$, and this is not a bounded function, though it is a Borel function. Here, measure theory helps, if the pvm has bounded support. The definition of the *support* of a pvm P is much the same as in regular measure theory:

$$\operatorname{Supp} P := \mathbb{R} \backslash \bigcup \{U \subset \mathbb{R} \mid P(U) = 0 \text{ and } U \text{ is open}\}.$$

The only, quite minor difference is that 0 in the condition $P(U) = 0$ refers to the projection operator 0. By definition, $\operatorname{Supp} P$ is a closed subset of $\mathbb{R}$, being the complement of an open subset. It also can not be empty, since if it were we would have that $I_{\mathcal{H}} = P(\mathbb{R}) = 0$, which is impossible since $\mathcal{H} \neq 0$.

By measure theory we always have

$$A = \int_{\operatorname{Supp} P} \lambda \, dP(\lambda).$$

So, if $\operatorname{Supp} P$ is a bounded subset of $\mathbb{R}$, then this integral exists, since the integrand *on this subset* is a bounded Borel function.

Here is where we see spectral theory entering the theory. First, we review some material from the functional analysis of Hilbert spaces.

Definition 2.3.2. Suppose that $A : \mathcal{D} \to \mathcal{H}$ is a linear operator defined on the dense subspace $\mathcal{D}$ of a Hilbert space $\mathcal{H}$. Then we define the *resolvent set* of A to be the following subset of the complex numbers:

$$\mathrm{Res}(A) := \{\lambda \in \mathbb{C} \,|\, \exists T_\lambda \in \mathcal{L}(\mathcal{H}) \text{ s.t. } (\lambda I - A)T_\lambda = I_{\mathcal{H}} \text{ and } T_\lambda(\lambda I - A) = id_{\mathcal{D}}\}.$$

One says λ is in the resolvent set provided that the (densely defined) operator $\lambda I - A$ has a globally defined, bounded inverse. Then the *spectrum* of A is defined as the complementary subset of $\mathbb{C}$, that is

$$\mathrm{Spec}(A) := \mathbb{C} \backslash \mathrm{Res}(A).$$

In functional analysis, one proves that for self-adjoint operators A we have that $\mathrm{Spec}(A)$ is a closed, non-empty subset of the real line $\mathbb{R}$ and, moreover, that $\mathrm{Spec}(A)$ is a bounded subset if and only if A is a bounded operator. (The diligent reader will have noticed that we have not defined what it means for a densely defined operator to be bounded. Even worse, we have not defined what it means for a densely defined operator to be self-adjoint. See your favorite functional analysis text for these details as well as the proof of the next theorem.)

Now, we come to a relation between the pvm and spectral theory.

Theorem 2.3.2. *Let A be a densely defined self-adjoint operator, and let P_A be the unique pvm associated to A by the spectral theorem. Then*

$$\mathrm{Supp}\, P_A = \mathrm{Spec}(A).$$

Putting these results together we see that for bounded self-adjoint operators the integral in (2.3.1) is well-defined. Of course, it still remains to prove the equation (2.3.1). As is typical in functional analysis there are a lot of technical details to deal with in order to give meaning to the integral in (2.3.1) for an unbounded self-adjoint operator A and, having done that, to prove the equation (2.3.1) holds.

2.4   The Finite Dimensional Case

Leaving those technical details to the texts, let us understand in detail what the Spectral Theorem says in the case when $\mathcal{H}$ is finite dimensional, since this is often not presented in a course of functional analysis. In that case we have a situation in linear algebra. The self-adjoint operator is then identified with an $n \times n$ Hermitian matrix A with $n \geq 1$. Since only the whole space $\mathcal{H}$ is a dense subspace (because $\dim \mathcal{H}$ is finite), the matrix A maps all of $\mathcal{H}$ to itself. In linear algebra one proves that the set of eigenvalues is a finite, non-empty subset of $\mathbb{R}$. Let $\mathrm{Spec}\, A = \{\lambda_1, \ldots, \lambda_k\}$ where $1 \leq k \leq n$, denote that subset of eigenvalues, that is these are the k *distinct* eigenvalues of A. The diagonalization theorem for A says that there is a basis of A with respect to which A has diagonal form with the eigenvalues appearing along the diagonal, each with a number of times equal to its multiplicity. Let's put this statement into another equivalent notation. For each $1 \leq j \leq k$ we define the *spectral subspace* associated with the eigenvalue λ_j by

$$V_j := \{\psi \in \mathcal{H} \mid A\psi = \lambda_j \psi\}.$$

Since λ_j is an eigenvalue, there is an eigenvector (non-zero by definition) in V_j. In short, $V_j \neq 0$. One proves for $i \neq j$ that V_i and V_j are orthogonal subspaces, since they correspond to eigenvalues $\lambda_i \neq \lambda_j$. The diagonalization of A is then realized by choosing a basis $\mathcal{B}_j$ of V_j for each $1 \leq j \leq k$ and proving that the union $\mathcal{B} := \bigcup_j \mathcal{B}_j$ is a basis of $\mathcal{H}$. Since A restricted to V_j is simply multiplication by λ_j, the matrix of A in the basis $\mathcal{B}$ is the sought for diagonal matrix representation of A. This means that we have an orthogonal decomposition

$$\mathcal{H} = V_1 \oplus \cdots \oplus V_k \tag{2.4.1}$$

such that A acts as multiplication by a scalar on each summand.

We now put this into the language of projections. For each spectral subspace V_j there is a unique projection such that $\mathrm{Ran}\, P_j = V_j$. The very fact that these are projections says that $P_j^* = P_j$ and $P_j^2 = P_j$. Moreover, the fact that V_i is orthogonal to V_j for all $i \neq j$ implies that $P_i P_j = 0$ for $i \neq j$. Note that this property is not shared by numbers (say, real or complex numbers as you wish) since in that case the product $ab = 0$ implies either $a = 0$ or $b = 0$. But these non-zero projections satisfy $P_i P_j = 0$ for $i \neq j$. In any event we use this convenient condensed notation: $P_i P_j = \delta_{ij} P_i$ for all $1 \leq i, j \leq k$, where δ_{ij} is the Kronecker delta. And we can write (2.4.1) in this notation as

$$I_{\mathcal{H}} = P_1 + \cdots + P_k,$$

which is called a *resolution of the identity*. This is not too exciting, since there are many, many resolutions of the identity. What makes this resolution of the identity interesting is that it comes from the diagonalization of A. In fact, since A acts as multiplication by the eigenvalue λ_j on the subspace

V_j, it follows quickly that

$$A = \lambda_1 P_1 + \cdots + \lambda_k P_k, \qquad (2.4.2)$$

which is called the *spectral representation* of the Hermitian matrix A. This is completely equivalent to the diagonalization of A. We next re-write the finite sum on the right side of (2.4.2) as an integral with respect to a pvm P. We clearly want P to satisfy $P(\{\lambda_j\}) = P_j$ for each $1 \leq j \leq k$. We also want $P(\mathbb{R}\backslash\mathrm{Spec}\, A) = 0$, the zero operator. If we can do that, then by standard identities of measure theory (only now applied to a pvm), we have

$$\int_{\mathbb{R}} \lambda\, dP(\lambda) = \int_{\mathbb{R}\backslash\mathrm{Spec}A} \lambda\, dP(\lambda) + \int_{\mathrm{Spec}A} \lambda\, dP(\lambda)$$

$$= 0 + \int_{\bigcup_{j=1}^{k}\{\lambda_j\}} \lambda\, dP(\lambda) = \sum_{j=1}^{k} \int_{\{\lambda_j\}} \lambda\, dP(\lambda) = \sum_{j=1}^{k} \lambda_j P(\{\lambda_j\})$$

$$= \lambda_1 P_1 + \cdots + \lambda_k P_k$$

as desired. So, it only remains to define the pvm with the required properties. But for every Borel subset $B \subset \mathbb{R}$ we simply define

$$P(B) := \sum_{j=1}^{k} \chi_B(\lambda_j)\, P_j,$$

where the characteristic function χ_S of any set S is defined as

$$\chi_S(\lambda) := \begin{cases} 1 & \text{if } \lambda \in S, \\ 0 & \text{if } \lambda \notin S. \end{cases}$$

An alternative way to write this is as

$$P(B) = \sum_{\{j\,|\,\lambda_j \in B\}} P_j,$$

that is, add up all the projections associated to the eigenvalues that lie in B.

One readily checks that P so defined is a pvm and that it satisfies the desired properties. The moral of this story is that the diagonalization of a Hermitian matrix is exactly the finite-dimensional special case of the Spectral Theorem. Another way to phrase this is to say that the Spectral Theorem generalizes the diagonalization of a Hermitian matrix to the infinite dimensional case. Note that in this generalization a finite sum has been replaced by an integral, which in certain cases will be a finite sum and in other cases will be an infinite sum. However, the complete generalization to the infinite dimensional case requires integrals and the corresponding measure theory with pvm's.

2.5 Quantum Probability: The First Steps

With all this mathematical background we are ready to see probability theory in the context of quantum theory. All the ways of introducing probability into quantum physics is what is called Quantum Probability.

The first thing to note is that a pvm is quite similar, though not identical, to a probability measure. This is seen for example in the normalization conditions. But even more impressive is that there is a partial order on self-adjoint operators, which we can restrict to the special case of projection operators (which are, as you will recall, self-adjoint). It turns out that for any projection P we have $0 \leq P \leq I_{\mathcal{H}}$, which are inequalities of self-adjoint operators. So projections lie between the two extreme projections possible. This is analogous to classical probability theory, where any probability p lies between the two extreme probabilities, namely $0 \leq p \leq 1$, which are inequalities of numbers.

To make the relation tighter with classical probability, we note that there is a converse of the Spectral Theorem.

Theorem 2.5.1 (Converse to Spectral Theorem). *Suppose that $\mathcal{H}$ is a Hilbert space and $P : \mathcal{B}(\mathbb{R}) \to \mathcal{L}(\mathcal{H})$ be a pvm. Then*

$$A := \int_{\mathbb{R}} \lambda \, dP(\lambda) \tag{2.5.1}$$

defines a self-adjoint operator, densely defined in $\mathcal{H}$. Moreover, the pvm P_A associated to A satisfies $P_A = P$.

Moreover, and more importantly, this result together with the Spectral Theorem gives a bijection between the set of all self-adjoint operators acting in $\mathcal{H}$ and the set of all pvm's taking values in $\mathcal{L}(\mathcal{H})$. One way the bijection sends the self-adjoint operator A to its pvm P_A. The other way a pvm P is sent to the self-adjoint operator given in (2.5.1). This is an amazing result, since it sets up a dictionary between objects of analysis (self-adjoint operators) and objects of measure theory (pvm's).

Therefore the second basic assumption in Section 2.2 of quantum theory, which says physical measurements are represented by self-adjoint operators, translates to saying that physical measurements are represented by pvm's. We are getting very close to classical probability theory using just this basic assumption of quantum theory. For the next step we consider a self-adjoint operator A (or equivalently, its pvm P_A) and a unit vector $\psi \in \mathcal{H}$, which represents a pure state, according to the first basic assumption of quantum theory in Section 2.2. Next, from these two mathematical objects, given by quantum physics, we define for every Borel subset $B \subset \mathbb{R}$ the expression on the left side here:

$$P_\psi(A \in B) := \langle \psi, P_A(B)\,\psi \rangle. \tag{2.5.2}$$

We could read the left side in full detail as follows: The probability that the measurement of A gives a value in B given that the quantum system is 'described' by the state ψ. A more colloquial reading without referencing any measurement is: The probability given ψ that A is in B. Of course, this terminology should be justified. The first result to prove is that the right side satisfies $0 \leq \langle \psi, P_A(B)\,\psi \rangle \leq 1$. The next thing to prove is that for the state ψ fixed and for the self-adjoint operator A fixed, the function defined for every Borel subset B of $\mathbb{R}$ by $B \mapsto \langle \psi, P_A(B)\,\psi \rangle$ is a classical probability measure on the real line $\mathbb{R}$, that is, it is σ-additive and satisfies the normalizations of a classical probability measure:

$$\langle \psi, P_A(\emptyset)\,\psi \rangle = 0 \quad \text{and} \quad \langle \psi, P_A(\mathbb{R})\,\psi \rangle = 1.$$

All of this follows immediately from the definitions and standard properties of Hilbert space. I hope the reader appreciates that the tricky bit in this theory is getting the definitions right. In functional analysis, one calls the probability measures in (2.5.2) the *spectral measures*. But in the physics community, (2.5.2) is known as *Born's Rule*, and although Max Born stated it differently it was worthy of his receiving the Nobel prize mainly for this achievement in quantum physics. Logically, it is fair to say that Born's Rule is the starting point of Quantum Probability.

Let us pause for an interlude to remark that there are some equivalent formulas for Born's Rule (2.5.2). Continuing with the notation established above and using standard results in Hilbert space theory we have

$$P_\psi(A \in B) = \langle \psi, P_A(B)\,\psi \rangle = \langle P_A(B)\psi, P_A(B)\,\psi \rangle = ||P_A(B)\,\psi||^2$$
$$= Tr(P_A(B)\,E_\psi) = Tr(E_\psi\,P_A(B))$$

where $E_\psi := |\psi\rangle\langle\psi|$ is a rank 1 density matrix (using Dirac notation) and Tr denotes the *trace* of a trace class operator. The last two formulas have the advantage of expressing the probability of Born's Rule in terms of the trace.

This is also a good time to note that there is something more general going on here. The event $P_A(B)$ is the formulas above can be replaced by any event E to give this definition of the *probability of a quantum event, given a pure state* ψ, as

$$P_\psi(E) := \langle \psi, E\psi \rangle = \langle E\psi, E\psi \rangle = ||E\psi||^2 = Tr(E_\psi\, E). \qquad (2.5.3)$$

The reader should note that this is a new meaning for the word 'probability'. Quantum events do not form a σ-algebra if $\dim \mathcal{H} \geq 2$, However, there are enough properties to justify this terminology. First, $0 \leq P_\psi(E) \leq 1$ since

$$0 \leq ||E\psi||^2 \leq ||E||^2_{op} ||\psi||^2 \leq 1$$

by (2.5.3) and basic Hilbert space theory.

Next, we have the normalizations $P_\psi(I) = 1$ and $P_\psi(0) = 0$, where the 0 on the left side of the last equation is the zero operator. This topic will be presented in more detail in Section 2.8.

And Definition (2.5.3) in turn is a particular case of the *expected value* of an observable $A = A^*$ with respect to ψ, which is defined as

$$\mathcal{E}_\psi(A) := \langle \psi, A\psi \rangle. \qquad (2.5.4)$$

A generalization is obtained by using the Definition (2.5.4) for *all* $A \in \mathcal{L}(\mathcal{H})$, and then the linear map $\mathcal{E}_\psi : \mathcal{L}(\mathcal{H}) \to \mathbb{C}$ so obtained gives an example of a *non-commutative integral*. The theory of non-commutative integration is found in many places in the literature.

Let's return to the main argument. So, by applying two basic assumptions of quantum theory in Section 2.2 (which really are widely accepted) we have arrived at an infinite number of classical probability measures for a given observable. One could say: This is a feature, not a bug. It only remains to understand what, if any, role these classical probability measures play in quantum physics. It is difficult to argue that they have no relevance or meaning in quantum physics, since they arise from two basic properties of quantum theory plus some (well, admittedly a lot of) mathematics. It is difficult to argue that there they are, but who cares? The standard way of dealing with this situation is to use these probability measures to save ('explain' if you wish) the phenomena of the relative frequencies that occur in the third basic property of quantum theory. However, do notice that our argument for arriving at these classical probability measures did not use the third basic assumption in Section 2.2.

It might appear that I have proved Axiom 4 in Chapter 1 from other axioms of quantum theory. That is not correct, since Axiom 4 is only a mathematical definition within the theory. However, it is a definition with a possible physical significance since it is related to observations. Using the probability measures in (2.5.2) to understand measured relative frequencies could be the physical significance of Axiom 4 as well as of the time dependent Born's Rule (1.6.1) in Axiom 5. But single event probability is not relevant, since only correlations of multiple events are relevant in science. This last assertion may seem unreasonable, since in standard textbook quantum theory the only probabilities considered are those of a single measurement, which is actually a type of event. But I maintain that the study of a single event (or measurement) without considering any other event (or measurement) is not what is done in scientific experiment and observation. One also always considers something else 'co-incident, both in time and space' as said by Alfred Russel Wallace. (See the Preface.)

However, two generally accepted basic assumptions of quantum theory have led us to the doorstep of Axiom 4. We just need an extra shove to cross the threshold to arrive at Axiom 4. This is quite different from Schrödinger's equation, though some authors put it on an equal footing with Born's Rule. But mathematical considerations alone do not get one close to Schrödinger's equation. One could just as well arrive at the Klein-Gordon equation or the Dirac equation. I consider Born's Rule as very nearly inevitable, but not so Schrödinger's equation. I have argued earlier that it is reasonable to replace Schrödinger's equation with Born's Rule as the fundamental time evolution equation of quantum theory. But we have yet to see the *Generalized Born's Rule*. The next section is a step in that direction.

2.6 Quantum Conditional Probability

The probability of an event occurring, given that another event has already occurred, is a standard topic in classical probability. This corresponds to the relative frequencies measured for such a pair of events. This is known as *conditional probability*. So, we now consider the quantum analogue of conditional probability. Recall that if $(\Omega, \mathcal{F}, P)$ is a classical *probability space* we define the conditional probability of an event $E_1 \in \mathcal{F}$ given the event $E_2 \in \mathcal{F}$ as

$$P(E_1|E_2) := \begin{cases} \dfrac{P(E_1 \cap E_2)}{P(E_2)} & \text{if } P(E_2) \neq 0 \\ 0 & \text{otherwise.} \end{cases} \qquad (2.6.1)$$

It is an elementary exercise to show that for a fixed event $E_2 \in \Omega$ that satisfies $P(E_2) \neq 0$ that P' defined for $E_1 \in \mathcal{F}$ by $P'(E_1) := P(E_1|E_2)$ is a probability measure on Ω. By a slight abuse of the terminology of C^*-algebras a probability measure is sometimes called a *state*.

We immediately have from (2.6.1) that $P(E_1|E_2)P(E_2) = P(E_1 \cap E_2)$ for all $E_1, E_2 \in \mathcal{F}$. From this we get Bayes' Theorem of classical probability:

$$P(E_1|E_2)P(E_2) = P(E_1 \cap E_2) = P(E_2 \cap E_1) = P(E_2|E_1)P(E_1). \quad (2.6.2)$$

We shall see that these simple formulas do not carry over to quantum probability. So, this classical probability theory serves as an analogy, nothing more.

Now, suppose that S, T are self-adjoint operators acting in $\mathcal{H}$ and that B, C are Borel subsets of $\mathbb{R}$. The probability that S has a value in B for a state ψ is given, as we have seen, by *Born's Rule:*

$$P_\psi(S \in B) = \langle \psi, P_S(B)\psi \rangle = ||P_S(B)\psi||^2. \quad (2.6.3)$$

We wish to define the conditional probability that $T \in C$ given that $S \in B$ for a state ψ. This will be denoted as $P_\psi(T \in C \,|\, S \in B)$. We assume that this has the form $P_{\psi_1}(T \in C)$ for some state ψ_1, that is to say, it will be a quantum probability of a single event. The implicit motivation for this assumption is that in standard quantum theory probability is only defined for single and not multiple events.

It is also reasonable to assume that the choice of ψ_1 should depend on the first event $S \in B$ and the state ψ. The usual textbook definition is

$$\psi_1 := \frac{P_S(B)\psi}{||P_S(B)\psi||} \quad (2.6.4)$$

provided $P_S(B)\psi \neq 0$. This definition has been given several unfortunate names. I will comment on those names and the significance of this definition after we have seen some of this material developed further. But first here is the important definition that this leads up to.

Definition 2.6.1. Let S, T be self-adjoint operators, B, C Borel subsets of $\mathbb{R}$ and ψ a unit vector. Then the (*quantum*) *conditional probability* that T has a value in C, given ψ and that S has a given value in B, is defined by

$$P_\psi(T \in C \,|\, S \in B) := P_{\psi_1}(T \in C) \quad \text{where } \psi_1 := \frac{P_S(B)\psi}{||P_S(B)\psi||} \quad (2.6.5)$$

if $P_S(B)\psi \neq 0$. Also, if $P_S(B)\psi = 0$ we define $P_\psi(T \in C \,|\, S \in B) := 0$.

Some might prefer to leave the expression $P_\psi(T \in C \mid S \in B)$ undefined in case $P_S(B)\psi = 0$. Taking S, T, B, ψ fixed and if also $P_S(B)\psi \neq 0$, then the mapping $C \mapsto P_\psi(T \in C \mid S \in B) = P_{\psi_1}(T \in C)$ defines a probability measure on $\mathbb{R}$. By using the formula in the next theorem one can see that the expression $P_\psi(T \in C \mid S \in B)$ is not even finitely additive in B.

Theorem 2.6.1. *With the notation of this definition we have that*

$$P_\psi(T \in C \mid S \in B) = \frac{||P_T(C)P_S(B)\psi||^2}{P_\psi(S \in B)} \tag{2.6.6}$$

provided that $P_\psi(S \in B) \neq 0$.

Proof. Expanding this definition out using the Definition (2.6.5) we see that the quantum conditional probability satisfies

$$P_\psi(T \in C \mid S \in B) = P_{\psi_1}(T \in C) = \left\langle \frac{P_S(B)\psi}{||P_S(B)\psi||}, P_T(C)\frac{P_S(B)\psi}{||P_S(B)\psi||} \right\rangle$$

$$= \frac{1}{||P_S(B)\psi||^2}\langle P_S(B)\psi, P_T(C)P_S(B)\psi \rangle$$

$$= \frac{||P_T(C)P_S(B)\psi||^2}{P_\psi(S \in B)}. \qquad \square$$

For doing computations the expression (2.6.6) is much more useful than Definition 2.6.5 and, in fact, is a preferable definition of quantum conditional probability, thereby avoiding the intermediate step (2.6.4) used to define ψ_1. Also, it immediately follows for *any* value of $P_S(B)\psi$ that

$$P_\psi(T \in C \mid S \in B)\, P_\psi(S \in B) = ||P_T(C)P_S(B)\psi||^2.$$

This definition of conditional probability immediately allows us to define independence. This topic is discussed in the next section in the more general context of quantum events.

Another immediate consequence of the definition is that for any $T = T^*$ and Borel subset C of $\mathbb{R}$ we have $P_\psi(T \in C \mid T \in C) = 1$ for all ψ satisfying $P_\psi(T \in C) \neq 0$. This should be compared with the equally trivial identity that holds in classical probability theory: $P(E|E) = 1$ for every event E, provided that $P(E) \neq 0$.

After seeing all this the reader should realize that the Definition (2.6.4) plays an intermediate, minor role and could have been avoided by taking (2.6.6) as the definition of conditional probability, as is done in [40]. But much ado has been spent on this so-called *collapse condition* (2.6.4).

It has also been dubbed the *quantum jump*, the *projection postulate* and the *collapse of the wave function*. So, much ink has been spilled to 'explain' or 'justify' this formula (often under the term 'interpretation') that I am obliged to make some comments about it.

The points of contact with observational data are the relevant, essential features of this, or any, theory. For quantum theory those are its quantum probabilities of sequences of multiple events. The well known Born's Rule (2.5.2) for single events is not relevant, since science is based on correlations. Rather the Generalized Born's Rule (2.10.2) given below is the appropriate theoretic expression for understanding sequences of multiple events. We can now see that this so-called collapse is not an event, but rather one step in a method for calculating the conditional probability of two sequential events that occur physically. So, collapse is not something that occurs. But if one thinks that it, rather than only the two events, is also what is happening, then one easily arrives at 'paradoxical' conclusions. For example, since no reference is made to spacetime coordinates in the definition of conditional probability (other than a time order in some frame of reference), one could think that 'something happens' instantaneously and non-locally, which is sometimes called 'spooky action at a distance'.

A similar situation holds in classical probability theory where the original state P is replaced as explained above by the state P' without any reference to spacetime coordinates. But in this situation the unfortunate terminology of collapse and its paradoxical consequences are not found in the literature.

When looking at different isomorphic models of quantum theory, it is perfectly acceptable to have discrepancies over matters that do not have anything to do with observations. What is relevant are the probabilities, not the specific manners for calculating them in a particular model. Conditional probability has the same formula in all models and produces the same value in all isomorphic models.

For example, in the Schrödinger model the states vary in time while the pvm's do not. On the contrary, in the Heisenberg model the opposite obtains: The pvm's vary in time while the states do not. What is the same in both are the probabilities for the same events. This is what makes these two models isomorphic. It is important to note that Born's Rule and its generalization are exactly the same formulas in both of these models. This is *covariance*. When the time dependence is introduced into these equations (but in different manners depending on the model chosen), the calculated time dependent probabilities are identically equal in these two models.

This is *invariance*. The collapse formula (2.6.4) does have a mathematical sense to it, that is, it is simply a mathematical operation. However, it is devoid of any significance as a physical event. Do be careful! The physical significance of the theory comes only later in the formula for the *one* quantum conditional probability (2.6.6) and the application of it to multiple quantum events.

2.7 Quantum Consecutive Probability

Experiment and theory indicate that a time ordered sequence of two or more observations should have an associated probability. Experiments measure the relative frequencies of such sequences. The relative frequency of just one single event is never measured. Science is based on correlations of events. Therefore, we see the importance of the following definition, which dates back at least to Wigner's 1963 paper [44]. So, this is often called *Wigner's Rule*, though as I have indicated earlier I prefer to call it a special case of the Generalized Born's Rule. However, I fear that this definition has been unduly neglected in the literature. But the literature is vast, and my reading of it is necessarily limited.

Definition 2.7.1. Let S, T be self-adjoint operators, B, C Borel subsets of $\mathbb{R}$ and ψ a unit vector. The *(quantum) consecutive probability* is defined as

$$P_\psi(S \in B, T \in C) := ||P_T(C)P_S(B)\psi||^2. \qquad (2.7.1)$$

The left side should be read as follows: The probability given the state ψ that S takes a value in B and *subsequently* T takes a value in C.

In [40] consecutive probability is defined first and conditional probability is defined in terms of it. Here, the order of definitions is reversed. These are the same definitions as in [40], but the motivating language is somewhat different. Of course, another motivation of this definition is that it allows us to write (2.6.6) in this form reminiscent of classical conditional probability:

$$P_\psi(T \in C \mid S \in B) = \frac{P_\psi(S \in B, T \in C)}{P_\psi(S \in B)} \quad \text{if } P_\psi(S \in B) \neq 0. \qquad (2.7.2)$$

Some authors have the previous definition only for the case of commuting operators. Some say it is not meaningful for non-commuting operators, while others react adversely to the fact that $P_T(C)P_S(B)$ is not necessarily an event and so should not have an associated probability. It is well known that in general there is no joint probability density function on $\mathbb{R}^2$ corresponding to the consecutive probability (2.7.1), but this does not negate its utility.

The expressions (2.6.6) and (2.7.1) are generalization of Born's Rule for calculating quantum probabilities. They are special cases of the *Generalized Born's Rule*, which will be presented later.

We showed in Section 1.7 that Born's Rule (2.6.3) is invariant if we change the states by $\psi \mapsto U_t \psi =: \psi'$ and the pvm's by $P_S \mapsto U_t P_S U_t^* =: P_{S'}$, where U_t is unitary for all $t \in \mathbb{R}$. We will now show that the conditional probability (2.6.6) and the Generalized Born's Rule (2.7.1) are also invariant under this transformation. Given this notation we now have

$$P_{\psi'}(S' \in B) = ||P_{S'}(B)\psi'||^2 = ||U_T P_S(B) U_T^* U_T \psi||^2$$
$$= ||P_S(B)\psi||^2 = P_\psi(S \in B).$$

So, for $P_{\psi'}(S' \in B) = P_\psi(S \in B) \neq 0$ the conditional probability (2.6.6) satisfies

$$P_{\psi'}(T' \in C \,|\, S' \in B) = \frac{||P_{T'}(C)P_{S'}(B)\psi'||^2}{P_{\psi'}(S' \in B)}$$
$$= \frac{||U_t P_T(C) U_t^* U_t P_S(B) U_t^* U_t \psi||^2}{P_\psi(S \in B)}$$
$$= \frac{||P_T(C)P_S(B)\psi||^2}{P_\psi(S \in B)} = P_\psi(T \in C | S \in B).$$

On the other hand, if $P_{\psi'}(S' \in B) = P_\psi(S \in B) = 0$, then

$$P_\psi(T \in C \,|\, S \in B) = 0 = P_{\psi'}(T' \in C \,|\, S' \in B).$$

This proves that the quantum conditional probability is invariant under this transformation. The proof that the Generalized Born's rule is also invariant is much the same. The importance of these invariances is that these quantum probabilities give the same results in all isomorphic quantum theories.

Note that the self-adjoint operators S and T commute (by definition!) if and only if every projection $P_S(B)$ commutes with every projection $P_T(C)$. So for, commuting S and T we have by (2.7.1) for all Borel sets $B, C \subset \mathbb{R}$ that

$$P_\psi(S \in B, T \in C) = P_\psi(T \in C, S \in B). \tag{2.7.3}$$

This says that *in this case* changing the temporal order of the two events does not change the consecutive probability. For such events simultaneity of the two events makes sense and gives the same consecutive probability (2.7.3).

However, it is not too difficult to find examples (thinking of spin $1/2$, say) of non-commuting self-adjoint operators S and T for which there exist Borel sets B, C and a state ψ satisfying

$$P_\psi(S \in B, T \in C) \neq P_\psi(T \in C, S \in B). \qquad (2.7.4)$$

This implies that for such S and T the time order of the two events must not include simultaneity. This contrasts with the case of commuting S and T where simultaneity is allowed.

Perhaps a word or two about the notion of 'time order' would be useful. I have not assumed Galilean invariance of this theory. Nor have I assumed Lorentz invariance. Either one of these invariances can be incorporated into quantum theory. Of course, these lead to more axioms of quantum theory. But in either case there is a well defined notion of invariance of time order of events with respect to the appropriate symmetry group. That is the notion of time order which should be applied here. I suppose that this is how time order should be understood in a quantum theory of gravitation, even though such a theory does not exist at the time of writing this. In a Galilean theory the time order is absolute and so trivially invariant, but in a Lorentzian theory it is only invariant for events that are light-like or time-like with respect to each other. (I am excluding time reversal as a symmetry, of course.) This motivates an extra axiom for Lorentz invariant quantum theories that two events that are space-like should commute.

This discussion has relevance to the well known–*and false*–assertion that 'the electron can be at two places at the same time'. Let Q represent the position operator for the electron. It is clear that Q commutes with itself. So, for any state ψ

$$P_\psi(Q \in B, Q \in C) = P_\psi(Q \in C, Q \in B) \qquad (2.7.5)$$

for any pair of Borel subset B, C of $\mathbb{R}$. This is important, since it includes simultaneous events, which is the case we want to consider. It also says that the probability of these simultaneous events makes sense in quantum theory. It is now just a matter of computing the probability in (2.7.5). So, we start that calculation as follows:

$$P_\psi(Q \in C, Q \in B) = ||P_Q(B)P_Q(C)\psi||^2 = ||P_Q(B \cap C)\psi||^2.$$

The last equality is a property of pvm's that is proved in functional analysis. The expression 'two places' I interpret as meaning that the sets B and C are disjoint, that is, $B \cap C = \emptyset$, the empty set. But, by the definition of pvm we have that $P_Q(\emptyset) = 0$. We conclude that the probability in (2.7.5) is 0 for disjoint Borel sets B, C and for *any* state ψ. So, much for the theory. What about observations? Well, I claim that there has never been an observation of an electron in two places at the same time. Of course, if such an observation were ever made, this aspect of quantum theory would have to be modified. But the principle of conservation of electric charge would also have to be modified. So, if there is anything 'strange' or 'mysterious' to understand here, I do not know what that might be. I also will not discuss the relation of this result with classical physics, because I do not wish to discuss classical physics at all in this treatise. As a final point let me note that this argument is quite general; it applies to any self-adjoint operator and any pair of disjoint Borel subsets of $\mathbb{R}$.

The inequality (2.7.4) is equivalent to

$$P_\psi(T \in C \,|\, S \in B)\, P_\psi(S \in B) \neq P_\psi(S \in B \,|\, T \in C)\, P_\psi(T \in C)$$

which says that the simplest generalization of Bayes' Theorem (2.6.2) to quantum probability does not hold.

However (anticipating notation such as (2.8.2)), in this approach for events E_1, E_2 and a state ρ we have that

$$P_\rho(E_2 \,|\, E_1)P_\rho(E_1) = P_\rho(E_1, E_2) = Tr\big((E_2 E_1)^* E_2 E_1 \rho\big) = Tr(E_1 E_2 E_1 \rho)$$

and also by interchanging E_1 and E_2 that

$$P_\rho(E_1 \,|\, E_2)P_\rho(E_2) = Tr(E_2 E_1 E_2 \rho).$$

It follows that

$$P_\rho(E_2 \,|\, E_1)P_\rho(E_1) = P_\rho(E_1 \,|\, E_2)P_\rho(E_2) - Tr(E_2 E_1 E_2 \rho) + Tr(E_1 E_2 E_1 \rho).$$

So, we call this identity the *quantum Bayes' theorem*. The first term on the right side is what we expect from the classical Bayes' theorem, while the other two terms are called *quantum interference terms*. When the events E_1 and E_2 commute, the quantum interference terms cancel out, thereby giving an identity closely analogous to the classical Bayes' Theorem (2.6.2). The point being made here is that saying something positive, such as the quantum Bayes' theorem holds, is preferable to saying something negative, such as the classical Bayes' formula does not hold. Of course, both are true.

Note that the inequality (2.7.4) also contrasts with the classical case where $P(E_1 \cap E_2) = P(E_2 \cap E_1)$. Moreover, (2.7.4) precludes defining a classical probability measure π on $\mathbb{R}^2$ such that S takes a value in B and T takes a value in C in any order is equal to $\pi(B \times C)$. Nonetheless, (2.7.1) has reasonable 'marginals', namely

$$P_\psi(S \in \mathbb{R}, T \in C) = ||P_T(C)P_S(\mathbb{R})\psi||^2 = ||P_T(C)\psi||^2 = P_\psi(T \in C),$$

since $P_S(\mathbb{R}) = I$, the identity map. Similarly,

$$P_\psi(S \in B, T \in \mathbb{R}) := ||P_T(\mathbb{R})P_S(B)\psi||^2 = ||P_S(B)\psi||^2 = P_\psi(S \in B).$$

On the other hand, taking S, T, B and ψ fixed, the map that sends the Borel subset C of $\mathbb{R}$ to $P_\psi(T \in C, S \in B)$ is a classical probability measure on $\mathbb{R}$ as already remarked before.

2.8 Quantum Probability of Two Events

The results of the previous section have been presented in terms of events associated to pvm's. So, we have considered events such as $S \in B = P_S(B)$ and so forth. But it is useful to express these results more abstractly in terms of arbitrary events. Here are the definitions. The first part of the next definition is Born's Rule for one event, given a pure state. The second part is Born's Rule for one event, given a density matrix; it is a generalization of the first part.

Definition 2.8.1. Let E be an event, ψ be a unit vector and ρ be a density matrix. Then the probability of E given ψ is defined by Born's Rule

$$P_\psi(E) := \langle \psi, E\psi \rangle = ||E\psi||^2. \tag{2.8.1}$$

Moreover, the probability of E given ρ is defined by

$$P_\rho(E) := Tr(E\rho). \tag{2.8.2}$$

In the sequel (2.8.2) will be regarded as a special case of the Generalized Born's Rule.

Let's recall the standard justification of (2.8.2) in terms of (2.8.1). So, we take an orthonormal basis ϕ_k which diagonalizes the density matrix ρ, say $\rho = \sum_k \lambda_k |\phi_k\rangle\langle\phi_k|$ where $0 \leq \lambda_k \leq 1$ and $\sum_k \lambda_k = 1$. Then we compute

$$P_\rho(E) = Tr(E\rho) = \sum_k \langle\phi_k, E\,\rho\,\phi_k\rangle = \sum_k \langle\phi_k, E\,\lambda_k\,\phi_k\rangle = \sum_k \lambda_k\,\langle\phi_k, E\,\phi_k\rangle$$

$$= \sum_k \lambda_k\,P_{\phi_k}(E).$$

In short, $P_{\sum_k \lambda_k |\phi_k\rangle\langle\phi_k|}(E) = \sum_k \lambda_k\,P_{\phi_k}(E)$. This exhibits $P_\rho(E)$ as an infinite convex combination of the component probabilities $P_{\phi_k}(E)$, each with its corresponding weight factor λ_k. An immediate consequence of the previous formula is $0 \leq P_\rho(E) \leq 1$. Also the following normalizations are easily shown and provide more justification for (2.8.2):

$$P_\rho(I) = 1 \quad \text{and} \quad P_\rho(0) = 0.$$

Finally, there is a form of σ-additivity which says that for any countable family $\{E_j \,|\, j \in \mathbb{N}\}$ of orthogonal events, meaning that $E_j E_k = 0$ for all $j \neq k$, we have $P_\rho(\vee_j E_j) = \sum_j P_\rho(E_j)$. To see this note that

$$P_\rho(\vee_j E_j) = Tr(\vee_j E_j \rho) = Tr\left(\sum_j E_j \rho\right) = \sum_j Tr(E_j \rho) = \sum_j P_\rho(E_j).$$

$$(2.8.3)$$

While this looks like the σ-countable condition for classical measures, it is quite different. Note that a necessary, but not sufficient, condition for the family $\{E_j\}$ to be orthogonal is that $\{E_j\}$ is a commutative family, a very restrictive condition. These facts motivate the next well known definition.

Definition 2.8.2. A (*quantum*) *probability* on the set $\mathcal{E}$ of quantum events of a von Neumann algebra is a function $\pi : \mathcal{E} \to [0,1]$ such that

- Normalizations: $\pi(0) = 0$ and $\pi(I) = 1$.
- σ-additivity: For any countable family $\{E_j \,|\, j \in \mathbb{N}\}$ of orthogonal events in $\mathcal{E}$ we have

$$\pi\left(\bigvee_j E_j\right) = \sum_j \pi(E_j).$$

It follows from the discussion above that for any density matrix ρ the function $E \mapsto P_\rho(E)$ is a probability on $\mathcal{E}$. The question of the converse arises, that is, whether all probabilities on $\mathcal{E} \subset \mathcal{V}$ have this form. The answer is given by Gleason's theorem in [17] and its generalization. But it is worth mentioning again that the probability of a single, isolated event is irrelevant for understanding physical systems, where the correlations among two or more events are what one considers.

The next definition gives two more versions of Born's Rule, but now for *two time ordered* events. It is based on (2.6.6) and (2.7.1) of the previous section.

Definition 2.8.3. Let E_1, E_2 be quantum events and ψ be a unit vector. Then the *(quantum) consecutive probability* of the event E_1 and then later the event E_2, given the state represented by ψ, is

$$P_\psi(E_1, E_2) := ||E_2 E_1 \psi||^2.$$

(Note that on the right side the earlier event E_1 goes on the right.)

The *(quantum) condition probability* of E_2, given that E_1 has occurred and given a state represented by ψ, is defined provided that $E_1 \psi \neq 0$ by

$$P_\psi(E_2 \mid E_1) := \frac{P_\psi(E_1, E_2)}{P_\psi(E_1)} = \frac{||E_2 E_1 \psi||^2}{||E_1 \psi||^2}.$$

If $E_1 \psi = 0$, then we define $P_\psi(E_2 \mid E_1) := 0$.

We say that the *event E_2 is (quantum) independent of the event E_1 in that order*, *given a unit vector* ψ, if $E_1 \psi \neq 0$ and $P_\psi(E_2 \mid E_1) = P_\psi(E_2)$ or more generally, by clearing the denominator, if

$$P_\psi(E_1, E_2) = P_\psi(E_1) P_\psi(E_2). \tag{2.8.4}$$

We say that the *event E_2 is entangled with the event E_1 in that order*, *given a unit vector* ψ, if they are not independent. This is called *entanglement* and is a new definition of this term.

If S and T are self-adjoint operators, then the *ordered pair* (S, T) is said to be *independent with respect to* ψ if for all Borel subsets B, C of $\mathbb{R}$ the event $S \in B$ is independent of the event $T \in C$ with respect to ψ, that is

$$P_\psi(S \in B, T \in C) = P_\psi(S \in B) \, P_\psi(T \in C).$$

Otherwise, the ordered pair (S, T) of observables is said to be *entangled*.

As we shall see in Chapter 3, the standard definition of entanglement in terms of tensor products is just a special case of lack of independence as defined here in a new way. Due to the non-commutativity of quantum theory quantum independence, unlike classical independence, is not necessarily a symmetric relation. Entanglement is similarly defined in Section 2.10 for three or more events as the lack of quantum probabilistic independence.

Clearly if E_1 and E_2 commute, then this is a symmetric relation, that is, E_1 is independent from E_2, given ψ, if and only if E_2 is independent from E_1, given ψ. Also $P_\psi(E_1, E_1) = P_\psi(E_1)$ trivially holds.

This definition is taking us ever deeper into non-commutative territory. Axiom 1 implies that there are non-commuting events in quantum theory. For each state we already have a probability (cp. Definition 2.8.2) defined on the set of quantum events $\mathcal{E}$ in $\mathcal{V}$, and $\mathcal{E}$ is not a σ-algebra. But now we have, given a state, a probability defined on *ordered pairs* of events. (Neither E_1 nor E_2 is associated to a specific time. We only require the time order that E_1 occurs first and then later E_2.) This is yet another step beyond the probability theory of Kolmogorov based on σ-algebras. The question arises as to what are the properties of quantum consecutive probability.

Throughout the following we let ψ be any unit vector. First, note that $0 \le P_\psi(E_1, E_2) \le 1$, since $||E_2 E_1||_{op} \le 1$. Also, we have for any event E the intuitively transparent formulas for the marginals

$$P_\psi(E, I) = P_\psi(E) \quad \text{and} \quad P_\psi(I, E) = P_\psi(E).$$

So, any event E and I are independent, and in both orders, since $P_\psi(I) = 1$. Moreover, at the other extreme we have the normalizations

$$P_\psi(E, 0) = 0 \quad \text{and} \quad P_\psi(0, E) = 0.$$

And so any event E and the 'never-YES' event 0 are independent (and again in both orders), since $P_\psi(0) = 0$.

Also, we have σ-additivity in the second event of the ordered pair of events. Let the event $F = \bigvee_j F_j$, where $\{F_j\}$ is a countable family of pairwise orthogonal events. Using $F_j F_k = \delta_{j,k} F_j$ we have for any event E that

$$P_\psi(E, F) = ||FE\psi||^2 = \left\| \left(\bigvee_j F_j \right) E\psi \right\|^2 = \left\| \sum_j F_j E\psi \right\|^2$$

$$= \left\langle \sum_j F_j E\psi, \sum_k F_k E\psi \right\rangle = \sum_j ||F_j E\psi||^2 = \sum_j P_\psi(E, F_j).$$

However, σ-additivity fails in general in the first event for consecutive probability. To see this we continue with the above notation and compute

$$P_\psi(F, E) = \|EF\,\psi\|^2 = \left\|E\left(\bigvee_j F_j\,\psi\right)\right\|^2$$

$$= \left\langle E\left(\bigvee_j F_j\,\psi\right), E\left(\bigvee_k F_k\,\psi\right)\right\rangle$$

$$= \sum_{j,k}\langle EF_j\,\psi, E\,F_k\,\psi\rangle$$

$$= \sum_j\langle EF_j\,\psi, EF_j\,\psi\rangle + \sum_{j\neq k}\langle EF_j\,\psi, EF_k\,\psi\rangle$$

$$= \sum_j P_\psi(F_j, E) + \sum_{j\neq k}\langle\psi, F_jEF_k\,\psi\rangle.$$

This identity is called *quantum σ-additivity*. The first summation is what one expects from σ-additivity. But the second summation is a typical quantum term, which already in the case $F = F_1 \vee F_2$ is producing interference with the first term. That is, the second sum, which is called an *interference term*, can either increase or decrease the first term. However, if E commutes with all of the events F_j (which already commute among themselves by orthogonality), then we have for each pair $j \neq k$ that

$$\langle\psi, F_jEF_k\,\psi\rangle = \langle\psi, EF_jF_k\,\psi\rangle = 0$$

by the orthogonality condition $F_jF_k = 0$. Therefore, in the commutative case this characteristic quantum interference term vanishes identically.

Also it is important to remark that this interference term arises from the rules for computing quantum probabilities and from nothing else. Of course, those rules are based on Hilbert space properties, particularly on the fact that the set of events is not a σ-algebra. Note, for example, that there is no particle/wave duality being invoked here. In fact, there is no wave equation. There is no Superposition Principle for solutions or for states. There is no mention of the Uncertainty Principle, of Complementarity or of the Measurement Problem. There is no so-called 'self-interference' of a particle with itself. Even Schrödinger's equation is absent from the derivation of this result. If this interference term is not intuitive for you, it means that quantum probability is not intuitive for you.

The phrase 'consecutive probability' is never even defined in classical probability theory for two good reasons. First, the order of events is not important. Second, the conjunction of two events A_1, A_2 in a σ-algebra is their intersection $A_1 \cap A_2$, which is again in the σ-algebra, that is, it is also an event. So, in classical probability the probability of two (and by induction any finite sequence of) events is itself the probability of just one event. This is not so in quantum probability. It is easy to construct events E_1, E_2 in $\mathcal{L}(\mathbb{C}^2)$ with $0 \neq E_1 E_2 \neq E_2 E_1 \neq 0$ and yet $E_1 \wedge E_2 = 0$, the 'never-YES' event. In this case neither $E_1 E_2$ nor $E_2 E_1$ is an event. Nonetheless, the probabilities of sequences $P_\psi(E_1, E_2)$ and $P_\psi(E_2, E_1)$ make good sense and are not equal in general. I find that this makes quantum probability more intuitive than classical probability, since it is common experience that some sequences of two events are more likely in one order rather than in the opposite order. For example, it is more probable that the Sun is blocked by clouds and then it rains rather than the reverse.

Example 2.8.1. Here is a general example. Let $\mathcal{H}_1$ and $\mathcal{H}_2$ be Hilbert spaces. Then define $\mathcal{H} := \mathcal{H}_1 \otimes \mathcal{H}_2$, the Hilbert space tensor product. Let F_1 (resp., F_2) be an event acting on $\mathcal{H}_1$ (resp., $\mathcal{H}_2$). Put $E_1 := F_1 \otimes I_2$, $E_2 := I_1 \otimes F_2$, where I_j is the identity map of $\mathcal{H}_j$ for $j = 1, 2$. Then E_1 and E_2 are commuting events, but that is not enough to have independence. We also have to choose appropriately a unit vector $\psi \in \mathcal{H}$. We choose $\psi = \psi_1 \otimes \psi_2$, where ψ_1 (resp., ψ_2) is a unit vector in $\mathcal{H}_1$ (resp., $\mathcal{H}_2$). Then it is an easy exercise to show that E_1 is independent of E_2 with respect to ψ. Of course, this example gets very interesting if we use a unit vector that does not factorize, namely an *entangled state*. We will come back to this in Chapter 3.

To generalize Definition 2.8.3 for the case of a density matrix, consider this expression for a unit vector ψ:

$$P_\psi(E_1, E_2) = ||E_2 E_1 \psi||^2 = \langle E_2 E_1 \psi, E_2 E_1 \psi \rangle = \langle \psi, (E_2 E_1)^* E_2 E_1 \psi \rangle$$
$$= \langle \psi, (E_2 E_1)^* E_2 E_1 E_\psi \psi \rangle = Tr((E_2 E_1)^* E_2 E_1 E_\psi)$$
$$= Tr(E_2 E_1 E_\psi E_1 E_2)$$

where $E_\psi := |\psi\rangle\langle\psi|$ is a rank 1 projection operator. So, using the same notation, the corresponding definition for a density matrix ρ is

$$P_\rho(E_1, E_2) := Tr((E_2 E_1)^* E_2 E_1 \rho) = Tr(E_2 E_1 \rho E_1 E_2), \qquad (2.8.5)$$

the *consecutive (conjunctive) probability* of first E_1 and later E_2, given ρ. Note that $(E_2E_1)^*E_2E_1\rho$ is a trace class operator, since ρ is. So, its trace in (2.8.5) is well defined. However, its trace norm need not be equal to its trace.

Also the *conditional probability* of E_2, given E_1 and ρ, is defined[2] by

$$P_\rho(E_2 \mid E_1) := \frac{P_\rho(E_1, E_2)}{P_\rho(E_1)} = \frac{Tr((E_2E_1)^*E_2E_1\rho)}{Tr(E_1\rho)} = \frac{Tr(E_2E_1\rho E_1E_2)}{Tr(E_1\rho)}$$

(2.8.6)

provided that $P_\rho(E_1) \neq 0$. The formula (2.8.6) is usually derived from Lüders Rule. (See [28].) Here, I have by-passed Lüders Rule and given (2.8.6) directly as a definition. One can easily manipulate (2.8.6) to arrive at the formula usually found in the literature for this conditional probability. For example see [8], where a uniqueness result for this formula is also found. Since I arrived at (2.8.6) without being aware of the literature (such as [8]), my notation is not standard.

Since these formulas may be unfamiliar in the non-commutative context, let's see a justification of (2.8.5). As before, we let ϕ_k be an orthonormal basis that diagonalizes ρ, that is, $\rho = \sum_k \lambda_k |\phi_k\rangle\langle\phi_k|$ where $0 \leq \lambda_k \leq 1$ and $\sum_k \lambda_k = 1$. Then we see that

$$P_\rho(E_1, E_2) = Tr((E_2E_1)^*E_2E_1\rho) = \sum_k \langle \phi_k, (E_2E_1)^*E_2E_1\rho\,\phi_k \rangle$$

$$= \sum_k \langle E_2E_1\phi_k, E_2E_1\lambda_k\,\phi_k \rangle = \sum_k \lambda_k \,||E_2E_1\phi_k||^2$$

$$= \sum_k \lambda_k\,P_{\phi_k}(E_1, E_2).$$

So, $P_{\sum_k \lambda_k |\phi_k\rangle\langle\phi_k|}(E_1, E_2) = \sum_k \lambda_k\,P_{\phi_k}(E_1, E_2)$, showing $P_\rho(E_1, E_2)$ as an infinite convex combination of the probabilities $P(E_1, E_2 \mid \phi_k)$, each with its corresponding weight factor λ_k. As before, these probabilities are easily shown to be real numbers in the interval $[0, 1]$.

As in the case of pure states, we define the *event E_2 to be independent of the event E_1 given a density matrix* ρ, if $P_\rho(E_1) \neq 0$ and

$$P_\rho(E_2 \mid E_1) = P_\rho(E_2).$$

---

[2]This definition appeared in [3] in 1981, and so (2.8.5) is implicitly found there, too.

By clearing the denominator, the more general form is

$$P_\rho(E_1, E_2) = P_\rho(E_1) P_\rho(E_2).$$

In general this is not a symmetric relation, though it is if E_1 and E_2 commute.

Example 2.8.2. The events $E_1 = F_1 \otimes I$ and $E_2 = I \otimes F_2$ of Example 2.8.1 are independent in either order with respect to any density matrix $\rho = \rho_1 \otimes \rho_2$, where ρ_1 (resp., ρ_2) is a density matrix acting on $\mathcal{H}_1$ (resp., $\mathcal{H}_2$). The details are left to the reader.

2.9 Collapse as Part of an Algorithm

In this section we take (2.8.6) as the definition of conditional probability. Rewriting that definition a bit in the notation used there, we have that

$$P_\rho(E_2 \mid E_1) := \frac{P_\rho(E_1, E_2)}{P_\rho(E_1)} = \frac{Tr((E_2 E_1)^* E_2 E_1 \rho)}{Tr(E_1 \rho)}$$

$$= \frac{Tr(E_1 E_2 E_2 E_1 \rho)}{Tr(E_1 \rho)} = \frac{Tr(E_2 E_1 \rho E_1)}{Tr(E_1 \rho)},$$

provided that $P_\rho(E_1) \neq 0$ which we will assume throughout this section. If we define

$$\rho' := \frac{E_1 \rho E_1}{Tr(E_1 \rho)}, \tag{2.9.1}$$

then, as is easily checked, ρ' is a trace class, positive operator. Its trace is

$$Tr(\rho') := \frac{Tr(E_1 \rho E_1)}{Tr(E_1 \rho)} = \frac{Tr(E_1^2 \rho)}{Tr(E_1 \rho)} = \frac{Tr(E_1 \rho)}{Tr(E_1 \rho)} = 1.$$

So, ρ' is a density matrix and hence a state. Also, we have

$$P_\rho(E_2 \mid E_1) = Tr(E_2 \rho') = P_{\rho'}(E_2).$$

So, we can compute the conditional probability on the left side using a two step algorithm. In the first step we use the first event E_1 and the given state ρ in order to compute ρ' using the Definition (2.9.1), which is actually *Lüders Rule* for the state ρ and the event E_1. (See [28].) Then in the second step we compute the probability of the second event E_2 using the 'new' state ρ'.

    This is a *two* step algorithm for calculating *one* conditional probability. The change from the state ρ to the state ρ' is a mathematical step in this algorithm. It does not correspond to something which happens physically (modeled theoretically as a quantum event), though it has colorfully been called 'collapse', especially in the case when ρ has rank 1. Since this step makes no reference to spacetime coordinates, it can easily be mistaken as something which happens non-locally and instantaneously; this has been referred to as 'spooky action at a distance'. This two step algorithm can not be presented as the computation of *one* conditional probability in texts where only single event probabilities are defined, but instead it is presented as these *two* separate calculations, the first of a 'collapse' given by Lüders Rule and the second of a one event probability. This then is a source of much confusion over how the second event 'knows' about the first event and about how the 'collapse' from ρ to ρ' can 'happen'. This confusion can be avoided by taking the Definition (2.8.6) of *one* conditional probability as basic and calculating directly with it and without any reference to 'collapse'.

Of course, there is also a simple three step algorithm for computing the conditional probability on the left side of (2.8.6). One considers any of the other expressions in (2.8.6) and computes first the numerator, second the denominator, and third their quotient. Just as with the two step algorithm given above, the third step corresponds to no physically occurring event. That is, there is no physical event which performs this arithmetical operation of division in the third step. If one were to think of this division as a physical event, one would have the problem of specifying where and when it occurs.

Let's see what this looks like in the particular case when $\rho = |\psi\rangle\langle\psi|$ for some unit vector ψ, since this is the case that is more likely to be familiar to the reader. Using identities and notation that we have already established, we have that

$$P_\psi(E_2 \mid E_1) := \frac{P_\psi(E_1, E_2)}{P_\psi(E_1)} = \frac{||E_2 E_1 \psi||^2}{||E_1 \psi||^2},$$

since $E_1\psi \neq 0$, follows from the assumption $P_\rho(E_1) \neq 0$ in this case.

If we define

$$\psi' := \frac{E_1\psi}{||E_1\psi||},$$

then $||\psi'|| = 1$ and $\rho' = |\psi'\rangle\langle\psi'|$ in this case. So, ψ' represents a state, which is called the 'collapse' by the event E_1 of the state represented by ψ. Also,

$$P_\psi(E_2 \mid E_1) = ||E_2\psi'||^2 = P_{\psi'}(E_2).$$

One again sees that *one* conditional probability is being computed in a *two* step algorithm, where the first step is the calculation of the new 'collapsed state' ψ' and the second step is the calculation of the one event probability of the second event E_2 using the new ψ'. All the further comments made above in the more general case clearly apply here, too.

2.10   Generalized Born's Rule with a State

Rather than spell out more details of the case of two events, we continue with the generalization to a finite sequence of time ordered events, which is now readily at hand. This updates Axiom 4.

Axiom 4 Updated: Time Independent Generalized Born's Rule with a State. Suppose that $E_1, E_2, \ldots, E_n$ for an integer $n \geq 1$ is an ordered sequence of events (possibly with repetitions) and let ψ be a unit vector. Then the *consecutive probability* that first E_1 occurs and then E_2 occurs and so on continuing until E_n occurs, given ψ, is defined as

$$P_\psi(E_1, E_2, \ldots, E_n) := ||E_n \cdots E_2 E_1 \psi||^2. \qquad (2.10.1)$$

Let Λ denote the *empty sequence* of events, not to be confused with $\emptyset$, the empty set. Note that Λ is vacuously ordered. Then we define $P_\psi(\Lambda) := 1$.

The *conditional probability* that a sequence of events $E_1, \ldots, E_n$ occurs in that time order given that the (possibly empty) sequence of events $F_1, \ldots, F_k$ has occurred previously and in that time order and given ψ, is defined as

$$P_\psi(E_1, \ldots, E_n \mid F_1, \ldots, F_k) := \frac{P_\psi(F_1, \ldots, F_k, E_1, \ldots, E_n)}{P_\psi(F_1, \ldots, F_k)}$$

$$= \frac{||E_n \cdots E_1 F_k \cdots F_1 \psi||^2}{||F_k \cdots F_1 \psi||^2}, \quad \text{provided the denominator is not zero.}$$

Otherwise we define $P_\psi(E_1, \ldots, E_n \mid F_1, \ldots, F_k) := 0$. These definitions are special cases of the following corresponding definitions, given the same sequences of events and a density matrix ρ.

The *consecutive probability* of the ordered sequence of events $E_1, \ldots, E_n$ for $n \geq 1$ is defined as

$$P_\rho(E_1, \ldots, E_n) := Tr\big((E_n \cdots E_1)^* E_n \cdots E_1 \rho\big).$$

Also, we define $P_\rho(\Lambda) := 1$. Let me remark again that consecutive probability is also called Wigner's Rule. See [44].

We say that a family of events $\{E_\alpha \,|\, \alpha \in A\}$ which has an order induced from a linear order on A is *independent* with respect to ρ if for every finite ordered subset $E_{\alpha_1}, E_{\alpha_2}, \ldots, E_{\alpha_n}$ for $n \geq 1$ with $\alpha_1 < \alpha_2 < \cdots < \alpha_n$ we have

$$P_\rho(E_{\alpha_1}, E_{\alpha_2}, \ldots, E_{\alpha_n}) = P_\rho(E_{\alpha_1})\, P_\rho(E_{\alpha_2}) \cdots P_\rho(E_{\alpha_n}).$$

If not, we say the family is *entangled*. This is a new definition of *entanglement*.

The ordered sequence $T_1, \ldots, T_n$ of self-adjoint operators is *independent* with respect to ρ if for every sequence of Borel subsets $B_1, \ldots, B_n$ of $\mathbb{R}$ the ordered sequence of events $T_1 \in B_1, \ldots, T_n \in B_n$ is independent with respect to ρ. Otherwise, we say that the operators are *entangled*.

With the same notation as above the *conditional probability* is defined as

$$P_\rho(E_1, \ldots, E_n \,|\, F_1, \ldots, F_k) := \frac{P_\rho(F_1, \ldots, F_k, E_1, \ldots, E_n)}{P_\rho(F_1, \ldots, F_k)} \qquad (2.10.2)$$

provided $P_\rho(F_1, \ldots, F_k) \neq 0$; otherwise it is defined to be 0.

All of the previous definitions of quantum probability are special cases of (2.10.2), which is called the *Generalized Born's Rule*. ∎

These probabilities are real numbers in the interval $[0, 1]$, but the language of σ-algebras has been left far behind. The time dependent version of Axiom 5 is now given by including the one-parameter groups E_t and S_t as before.

The ordering here of events reflects the time order of their occurrences, but does not associate them to specific times. However, since events are self-adjoint operators, they can have a time evolution in some models (such as the Heisenberg model). But this evolution does not impact the time order of these events.

Consider the union $\{F_1, \ldots, F_k, E_1, \ldots, E_n\}$ of the ordered sequences as a new ordered sequence of events. If this new ordered sequence is independent, then *necessarily* we have that

$$P_\rho(E_1, \ldots, E_n \,|\, F_1, \ldots, F_k) = P_\rho(E_1, \ldots, E_n)\, P_\rho(F_1, \ldots, F_k).$$

But this is not a *sufficient* condition for independence of the new ordered sequence, except in the case when $n = k = 1$.

Having defined independence of an ordered sequence of observables (which are quantum random variables), it is desirable to define identically distributed observables as well. Then we will be able to speak of an ordered sequence of independent, identically distributed (iid) observables.

Definition 2.10.1. Let S and T be self-adjoint operators. Then we say that S and T are *identically distributed* with respect to a density matrix ρ if $P_\rho(S \in B) = P_\rho(T \in B)$ for all Borel subsets B of $\mathbb{R}$. If they are also independent in some order, then we say they are *independent, identically distributed (iid)* observables in that order.

The construction of a finite sequence of iid observables can be done easily using tensor products.

    Everything in the formulas of these definitions is well known to anyone working in classical probability where each sequence of events becomes just one event. What is really new is not so much generalizing to sequences of non-commuting events, but rather identifying all of this as the Generalized Born's Rule (2.10.2) of quantum theory. Moreover, (2.10.2) is the fundamental time evolution equation of quantum theory, provided that the secondary, model dependent time evolutions are given for the set of events and for the set of states.

    The normalization conditions $P_\psi(\Lambda) = 1$ and $P_\rho(\Lambda) = 1$ could seem non-intuitive. In their defense they make the conditional probabilities work out when $k = 0$. But they seem to say that the probability that nothing happens is 1. I suspect that this is a trap of language. Think of ever longer sequences of events. Easily such very long sequences can have probability 0 or very near 0. But then thinking of starting with such a long sequence of events with small probability and then considering shorter and shorter contiguous sub-sequences of it. The intuition is that the probability increases. And when one arrives at the empty sequence of events, one then has the most probable situation. In more mundane terms we can say that events impose restrictions on probability and that, by removing all such restrictions, one gets the most probable outcome, that is, probability 1.

    Continuing with the intuition in the previous paragraph, we can define the probability of an infinite sequence of events.

Definition 2.10.2. Suppose $\{E_j \mid j \in \mathbb{N}\}$ is an infinite sequence of linearly ordered events with E_i preceding E_j if and only if $i < j$. Let ρ be a density matrix. Then we define the probability of the ordered sequence, given ρ, as

$$P_\rho(\{E_j\}) := \lim_{n\to\infty} P_\rho(E_1, \ldots, E_n) = \lim_{n\to\infty} Tr\big((E_n \cdots E_1)^* E_n \cdots E_1 \rho\big)$$

$$= \inf_n Tr\big((E_n \cdots E_1)^* E_n \cdots E_1 \rho\big).$$

If this limit exists, then clearly $0 \le P_\rho(\{E_j\}) \le 1$. It also satisfies obvious normalization conditions if all $E_j = I$ or if any $E_j = 0$.

The limit in this definition exists since the sequence is both bounded below by 0 and non-increasing. To prove the latter statement we compare the $(n+1)$–st term with the n–th term. As usual we write $\rho = \sum_k \lambda_k |\phi_k\rangle\langle\phi_k|$ where $0 \le \lambda_k \le 1$, $\sum_k \lambda_k = 1$ and $\{\phi_k\}$ is an orthonormal basis of $\mathcal{H}$ which diagonalizes ρ. The result is trivially true if $E_{n+1} = 0$. So, we assume that $E_{n+1} \ne 0$ which implies that $||E_{n+1}||_{op} = 1$. Then we have

$$Tr\big((E_{n+1}\cdots E_1)^* E_{n+1}\cdots E_1\rho\big)$$

$$= \sum_k \big\langle \phi_k, (E_{n+1}\cdots E_1)^* E_{n+1}\cdots E_1\rho\phi_k \big\rangle$$

$$= \sum_k \big\langle E_{n+1}\cdots E_1\phi_k, E_{n+1}\cdots E_1\rho\phi_k \big\rangle$$

$$= \sum_k \lambda_k \big\langle E_{n+1}\cdots E_1\phi_k, E_{n+1}\cdots E_1\phi_k \big\rangle$$

$$= \sum_k \lambda_k ||E_{n+1}E_n\cdots E_1\phi_k||^2 \le \sum_k \lambda_k ||E_{n+1}||_{op}^2 \, ||E_n\cdots E_1\phi_k||^2$$

$$= \sum_k \lambda_k ||E_n\cdots E_1\phi_k||^2 = \sum_k \lambda_k \big\langle E_n\cdots E_1\phi_k, E_n\cdots E_1\phi_k \big\rangle$$

$$= \sum_k \big\langle E_n\cdots E_1\phi_k, E_n\cdots E_1\lambda_k\phi_k \big\rangle = \sum_k \big\langle \phi_k, (E_n\cdots E_1)^* E_n\cdots E_1\rho\phi_k \big\rangle$$

$$= Tr\big((E_n\cdots E_1)^* E_n\cdots E_1\rho\big).$$

This proves that the sequence is non-increasing. And this fact is behind the assertion that the conditional probabilities are ≤ 1. However, this result does *not* mean that more events lowers the probability. It only says that adding more events *after* a given sequence of events lowers the probability. It is well known one can find events E_1 and E_2 acting on $\mathbb{C}^2$ such that $E_2 E_1 = 0$ and so, in particular, $P_\psi(E_2 E_1) = 0$, but that there exists an event F satisfying $P_\psi(E_2 F E_1) > 0$. (Think about light polarizing filters.)

Since E_1 is the first event and E_n is the last event, the expression $E_n\cdots E_1$ is well-known in quantum field theory. It is called a *time-ordered product*. Clearly, the actual calculation of these probabilities can be rather challenging in practice. Such probabilities could be difficult to check in the laboratory as well. One typically prefers experimental situations with few events in play. However, nature does not always smile favorably on the experimental scientist. Even if one wished to study just two consecutive events, there may be other uncontrolled intermediate events so that one is studying a situation with many events instead of just two. Such undesired intermediate events are pejoratively dubbed as *noise* (as if they were not physical phenomena which one could study) and the experimenter then works hard to eliminate them or, at least, to minimize their collective impact. Neither at an experimental level nor a theoretical level is there any 'mystery' about such noise that needs special explanation. It is simply very annoying. But to their merit some physicists do try to study this noise which they rename as *decoherence*.

It is important to remark, and trivial to verify, that the probabilities in Definition 2.10 are invariant under isomorphisms between models. So, it makes sense to suppose these probabilities could have physical significance. The Generalized Born's Rule also applies to the special case of measurements, which involve simply events having the form $T \in B$, where B is a small interval in $\mathbb{R}$ and T is a self-adjoint operator.

2.11 Generalized Born's Rule with no State

It makes mathematical sense to drop the state from the formulas of the last section and arrive at a definition of probability for sequences of events with no mention of states. This requires modifying Axiom 5 so that only events can be time dependent. Consequently the Schrödinger and interaction models now do not apply. This clashes with ideas from classical single event probability although it has some physical motivation, since this counter-intuitive and controversial definition is used to correctly discuss entanglement in Chapter 3. And so it does have relevance to physics.

Definition 2.11.1. Suppose that $E_1, E_2, \ldots, E_n$ for an integer $n \geq 1$ is an ordered sequence of events. Then the *consecutive (conjunctive) probability (with no state)* that first E_1 occurs and then E_2 occurs and so on continuing until E_n occurs is defined as

$$P(E_1, E_2, \ldots, E_n) := ||E_n \cdots E_2 E_1||^2_{op}. \qquad (2.11.1)$$

Let Λ denote the *empty sequence* of events. Then we define $P(\Lambda) := 1$.

The *conditional probability (with no state)* that the sequence of events $E_1, \ldots, E_n$ occurs in that time order given the sequence of events $F_1, \ldots, F_k$ having already occurred in that time order is defined as

$$P(E_1, \ldots, E_n \mid F_1, \ldots, F_k) := \frac{P(F_1, \ldots, F_k, E_1, \ldots, E_n)}{P(F_1, \ldots, F_k)}$$

$$= \frac{\|E_n \cdots E_1 F_k \cdots F_1\|_{op}^2}{\|F_k \cdots F_1\|_{op}^2} \quad \text{provided the denominator is not zero.}$$

(The definition on the first line is for all integers $k \geq 0$, while the second line only holds for $k \geq 1$.) Otherwise we define $P(E_1, \ldots, E_n \mid F_1, \ldots, F_k) := 0$.

Independence and entanglement with no state are defined with the same formulas as in Section 2.10, but with the state omitted.

Note that these probabilities are real numbers in the interval $[0, 1]$. This definition could appear to be non-intuitive, especially because of the following puzzling particular case. If $E \neq 0$ is an event, then $P(E) = \|E\|_{op}^2 = 1$.

This contradicts the intuition that single events should have non-trivial probabilities. However, this intuition is wrong. The probability of a single event without any relation to any other event has no physical significance. No experiment deals with just single events. Science is based on correlations among events. This is just fine, since we only need a theory for probabilities of two or more events. So, it just happens that there is a formula for the probability of a single event in this theory, but it is irrelevant.

In any case in Chapter 3 on Entanglement at first only the conditional probability with a state will be used although a similar analysis with the less intuitive conditional probability with no state will be discussed as well in order to show that this does make sense in an application.

This probability with no state contrasts sharply with the probability of an event, given a state ψ, for which we have that $P_\psi(E) = \|E\psi\|^2$ can be any number in the interval $[0, 1]$ for events $0 \neq E \neq I$. This accords with the classical idea that the probability of an event reflects to some degree the state in which the system 'finds' itself. But this is misleading language. When one says that the system 'finds' itself in a state, what one means invariably is that some other event (or events) precedes the event of interest. And so one is considering a conditional probability.

Here is a situation where the probability with no state makes total sense. This is an identity that we will use later. We let $\psi \in \mathcal{H}$ be a unit vector and let E be an event or a product of events. Then we have

$$P(E\,|\psi\rangle\langle\psi|) = ||\,E\,|\psi\rangle\langle\psi|\,||^2_{op} = \sup_{||\phi||=1} ||E\,|\psi\rangle\langle\psi|\,\phi||^2$$

$$= \sup_{||\phi||=1} |\langle\psi,\phi\rangle|^2||E\psi||^2 = ||E\psi||^2 = P_\psi(E). \qquad (2.11.2)$$

Let's see how this works in a familiar example. Let Q denote the position operator defined in the Hilbert space $L^2(\mathbb{R})$. Let $a, b \in \mathbb{R}$ satisfy $a < b$. Then the quantum event

$$Q \in [a, b] = P_Q\big([a, b]\big) \neq 0$$

and so $P(Q \in [a, b]) = 1$. This can be thought as a way of saying that

$$P_\psi(Q \in [a, b]) = ||\big(Q \in [a, b]\big)\psi||^2$$

can assume any value in $[0, 1]$, depending on the choice of the unit vector ψ. However, the quantum event

$$Q \in [a] = P_Q\big([a]\big) = 0$$

and so $P(Q \in [a]) = 0$. This is a way of saying in the language of quantum probability that the position observable can not give sharp numerical values, but can give values in non-trivial intervals.

2.12   Probability Amplitudes

In many formulations of quantum theory it is emphasized that the quantum probabilities are calculated as the absolute value squared of a probability amplitude. That is implicit in this approach in the case of a Type I factor when we have two unit vectors $\psi, \phi \in \mathcal{H}$, in which case we define their probability amplitude to be $A(\phi, \psi) := \langle\phi, \psi\rangle$. We introduce the notation $E_\phi = |\phi\rangle\langle\phi|$, the event that '$\phi$ occurs'. The probability of E_ϕ, given ψ, is

$$P_\psi(E_\phi) = \langle\psi, E_\phi\psi\rangle = \langle\psi, |\phi\rangle\langle\phi|\,\psi\rangle = \langle\psi, \langle\phi, \psi\rangle\phi\rangle = |\langle\psi, \phi\rangle|^2 = |A(\phi, \psi)|^2.$$

So, our axiomatization of Born's Rule gives the usual physical significance to the expression $|\langle\psi, \phi\rangle|^2$, although the inner product $\langle\psi, \phi\rangle$ itself has no physical significance.

For fixed ϕ the map $\psi \mapsto A(\phi, \psi)$ is linear provided that ψ is allowed to be *any* vector in $\mathcal{H}$, but that $\psi \mapsto |A(\phi, \psi)|^2$ is not linear. On the other hand, suppose that the state $\psi = \sum_k \lambda_k \phi_k$, where $\{\phi_k\}$ is an orthonormal set and the complex numbers λ_k satisfy $\sum_k |\lambda_k|^2 = 1$. Then we have a form of convexity of amplitudes: $A(\phi, \psi) = \sum_k \lambda_k \langle \phi, \phi_k \rangle = \sum_k \lambda_k A(\phi, \phi_k)$. But in general $|A(\phi, \psi)|^2$ and $\sum_k |\lambda_k \langle \phi, \phi_k \rangle|^2 = \sum_k |\lambda_k|^2 |A(\phi, \phi_k)|^2$ are not equal due to the well known interference terms in the expansion of $|A(\phi, \psi)|^2$.

Probability amplitudes can also be defined for all the other examples of quantum probability, but they are elements in the Hilbert space instead of being complex numbers. For example, for a quantum event E and a unit vector ψ we have $P_\psi(E) = ||A(E, \psi)||^2$, where we define the *probability amplitude* of the event E in the state ψ as $A(E, \psi) := E\psi \in \mathcal{H}$.

Essentially, probability amplitudes add more notation to the theory but without shedding much more light on it. However, that language is available if one wishes to use it.

2.13   Quantum Integrals

Having defined and studied quantum probability, it is now straightforward to define and study the quantum theory of integration as was mentioned earlier as $\mathcal{E}_\rho(A) := Tr(A\rho)$ for any $A \in \mathcal{L}(\mathcal{H})$ and density matrix ρ. We define this to be the *(non-commutative) integral of A with respect to the state ρ*. Using terminology from classical probability theory we can call this the *expectation of A with respect to ρ*. In quantum physics if $A = A^*$ one says that this is the *expected value of A in the state ρ*; this is well known since the early days of quantum theory, although only after Born's seminal paper [5] appeared. So, the next definition seems to be natural, though its importance is not clear.

Definition 2.13.1. Let $A_1, A_2, \ldots, A_n$ be an *ordered* sequence in $\mathcal{L}(\mathcal{H})$ and let ρ be a density matrix. We define the *time-ordered integral* of this sequence of operators with respect to ρ to be

$$\mathcal{E}_\rho(A_1, A_2, \ldots, A_n) := Tr(A_n \cdots A_2 A_1 \rho). \qquad (2.13.1)$$

This is also called the *expectation of the ordered sequence $A_1, A_2, \ldots, A_n$ with respect to ρ*.

Of course, the time ordered integral of an ordered sequence of operators is equal to the integral of a single operator, since

$$\mathcal{E}_\rho(A_1, A_2, \ldots, A_n) = \mathcal{E}_\rho(A_n \cdots A_2 A_1).$$

But the point is that non-commutativity of $\mathcal{L}(\mathcal{H})$ makes the order of the operators important. And that is what is underlining this definition.

The corresponding definition in usual measure theory, which is a commutative integration theory, would not have such importance.

There is a temptation to say that (2.13.1) is a 'state' that is associated to the probability for ordered sequences of events. However, it has some properties that argue against being so named. For example, for an ordered sequence $E_1, E_2, \ldots, E_n$ of events with $n \geq 2$ we have in general that

$$\mathcal{E}_\rho(E_1, E_2, \ldots, E_n) \neq P_\rho(E_1, E_2, \ldots, E_n).$$

So, with this definition the expectation does not extend the probability. Also, there is no apparent positivity property. One 'nice' property that it does have is the normalization $\mathcal{E}_\rho(I, I, \ldots, I) = Tr\,\rho = 1$. It also has reasonable marginals such as $\mathcal{E}_\rho(I, E_2, \ldots, E_n) = \mathcal{E}_\rho(E_2, \ldots, E_n)$ and so forth.

But now we also have quantum conditional probability at our disposal, and so we should have quantum conditional expectation as well.

Definition 2.13.2. Let $E_1, \ldots, E_k$ be an *ordered* sequence of events in $\mathcal{L}(\mathcal{H})$ and $A_1, \ldots, A_n$ be an *ordered* sequence of operators in $\mathcal{L}(\mathcal{H})$. Also let ρ be a density matrix. Then the *quantum conditional expectation* is defined as

$$\mathcal{E}_\rho(A_1, \ldots, A_n \mid E_1, \ldots, E_k) := \frac{\mathcal{E}_\rho(A_1, \ldots, A_n, E_1, \ldots, E_k)}{\mathcal{E}_\rho(E_1, \ldots, E_k)} \qquad (2.13.2)$$

provided $\mathcal{E}_\rho(E_1, \ldots, E_k) \neq 0$. (NB: The denominator is not $P_\rho(E_1, \ldots, E_k)$.)

Note in the Schrödinger, Heisenberg and interaction models that the time evolution extends naturally to $\mathcal{L}(\mathcal{H})$. Consequently, given a Hamiltonian these integrals are also time dependent in those models. This definition is quite different from the partial trace, which is a non-commutative sort of conditional expectation in the context of tensor product Hilbert spaces.

2.14   Born's Rule Redux

Until the end of the last section, I had been rather cavalier in using the term *Born's Rule*. What I mean by it is any formula in quantum theory that is a special case of (2.10.2). Since M. Born was the first to give such a formula in quantum theory, I have decided to credit him by calling (2.10.2) and its immediate consequences the *Generalized Born's Rule*. Actually, I have not yet presented Born's Rule in a more usual form. It might be instructive for the reader to see this in detail. To do this I shall dive ever so shallowly into historical waters.

The time independent version of Schrödinger's equation $H\psi = E\psi$ for a self-adjoint Hamiltonian operator $H = H^*$ is an eigenvalue problem with two unknowns for which it must be solved: the eigenvalue $E \in \mathbb{R}$ and its corresponding non-zero eigenvector ψ. Already in Schrödinger's first paper [36] on the subject it was noted that E represents an energy, but the physical significance of ψ was left unresolved in that paper. However, it seemed reasonable that the solution ψ should also have some physical significance. And a similar concern arises with the solution ψ_t of the time dependent Schrödinger equation.

In modern terminology M. Born addressed this in the specific case that $\psi \in L^2(\mathbb{R}^3)$. However, to avoid a lot of sub-indices let's consider the case $\psi \in L^2(\mathbb{R})$, since the same ideas apply. So, $\mathcal{H} = L^2(\mathbb{R})$ is the Hilbert space for this situation. The basic assumption is that the position of the system is a relevant observable, that is, the values of its pvm lie in the von Neumann algebra of the system. Here, the self-adjoint position operator $Q : D(Q) \to L^2(\mathbb{R})$ is defined on the dense subspace

$$D(Q) := \{\psi \in L^2(\mathbb{R}) \,|\, x\psi(x) \in L^2(\mathbb{R})\}$$

by the formula $Q\psi(x) := x\psi(x)$. But more importantly, the pvm of Q is $P_Q(B)\,\phi = \chi_B\,\phi$ for all Borel subsets B of $\mathbb{R}$ and all $\phi \in L^2(\mathbb{R})$. Here, χ_B is the characteristic function of the Borel set B, defined for all $x \in \mathbb{R}$ as

$$\chi_B(x) := \begin{cases} 1 & \text{if } x \in B, \\ 0 & \text{if } x \notin B. \end{cases}$$

All of these results about Q come from functional analysis. We continue by using Born's Rule as given in Axiom 4 to calculate the probability that the position of the system is in a Borel subset B of $\mathbb{R}$ given a unit vector $\psi \in L^2(\mathbb{R})$ as follows:

$$P_\psi(Q \in B) = \langle \psi, P_Q(B)\psi \rangle = \int_{\mathbb{R}} \psi(x)^* \big(P_Q(B)\psi(x)\big)\, dx$$

$$= \int_{\mathbb{R}} \psi(x)^* \chi_B(x)\psi(x)\, dx = \int_B \psi(x)^* \psi(x)\, dx = \int_B |\psi(x)|^2\, dx.$$

The expression $|\psi(x)|^2$ on the right here is the formula given by Born for the probability density for the position of the system given the unit vector ψ.

Of course, Born came to this conclusion without using all the tools of quantum probability, which came later. Actually, Born initiated the field of quantum probability by indicating the physical significance of this expression. This formula for the probability is one way of viewing the physical significance of the solution $\psi \in L^2(\mathbb{R})$ of the eigenvalue problem $H\psi = E\psi$. If we consider $P_\psi(T \in B)$ for some other self-adjoint operator T, we can give ψ some other physical significance. I do not wish to elevate this last comment to the level of a general principle of complementarity; it is merely another application of Born's Rule as given in Axiom 4.

Concerning a solution $\psi_t \in L^2(\mathbb{R})$ of the time dependent Schrödinger's equation, Born's Rule asserts that

$$P_{\psi_t}(Q \in B) = \int_B |\psi_t(x)|^2\, dx \qquad (2.14.1)$$

is the probability at every time $t \in \mathbb{R}$ that the position of the system is in the Borel set B. So, we see time dependent probability in quantum theory in this simple example. However, in this treatise we take equation (2.14.1) to be the fundamental time evolution equation of the position observable Q. Moreover, we see in this example that the time dependent Schrödinger's equation does not play any role in understanding the physical significance of ψ_t.

2.15   Comparison with Classical Probability

This section, just as the rest of this treatise, does not address the issues of comparing quantum theory with classical physical theory. Rather, it is a comparison of Kolmogorov's formulation in [25] of classical probability in terms of measure theory with the quantum theory of probability that has been presented here.

A key difference is the mathematical structure of the set of events. In classical probability the events are elements of a *σ-algebra* $\mathcal{F}$ whose elements are subsets of a non-empty *sample space* Ω. In particular this is a Boolean algebra. This means the various rules of Boolean algebra hold including the de Morgan identities. Another way of saying this is that the events obey the rules of classical logic that go back at least to the works of Aristotle. Thinking that events tell us that nature has certain properties, this means that those properties also satisfy the rules of classical logic.

In quantum theory the events are the closed subspaces ($\equiv$ projections) of a *complex* Hilbert space $\mathcal{H}$. (Note that the role of the complex numbers seems to be essential.) These form a complete orthomodular lattice for which the de Morgan identities fail if $\dim \mathcal{H} \geq 2$. Consequently, if one assigns 'properties' to these events, then classical logic will not apply to them. So, we must think differently about quantum events. Of course, we can always say that the event itself is a property, but this is quite distinct from how the properties of classical logic are structured.

Quantum events when viewed as projections lie inside a larger structure, namely the complex vector space $\mathcal{L}(\mathcal{H})$. To draw comparisons it is convenient to embed classical events in a larger structure, namely the complex vector space $\mathcal{M}(\Omega) := \{X : \Omega \to \mathbb{C} \mid X \text{ is measurable}\}$. Then a classical event $A \in \mathcal{F}$ is associated with its characteristic function $\chi_A \in \mathcal{M}(\Omega)$. One has $\chi_A^2 = \chi_A = \chi_A^*$. Conversely, for every $X \in \mathcal{M}(\Omega)$ satisfying $X^2 = X = X^*$, there exists a unique set $A \in \mathcal{F}$ such that $X = \chi_A$. So, we can define classical events equivalently as those $\chi \in \mathcal{M}(\Omega)$ satisfying $\chi^2 = \chi = \chi^*$. This compares favorably with the definition of a quantum event as those $E \in \mathcal{L}(\mathcal{H})$ satisfying $E^2 = E = E^*$.

In classical probability the observables are called *random variables* and are defined as those elements of $\mathcal{M}(\Omega)$ that are real valued, that is, those $X \in \mathcal{M}(\Omega)$ satisfying $X = X^*$. The *essential range* of any $X \in \mathcal{M}(\Omega)$ is the *spectrum* of X, denoted $\mathrm{Spec}(X)$. (See a text on measure theory for definitions.) If X is a random variable, then the elements of $\mathrm{Spec}(X)$ form a non-empty subset of $\mathbb{R}$, and its elements correspond to the values seen in the observations associated to X. For each Borel subset B of $\mathbb{R}$ there is an event in $\mathcal{F}$ that is denoted as $X \in B$ and is called the event that X is observed to have a value in B. It is defined as $X \in B := X^{-1}(B)$. If a probability measure P on $\mathcal{F}$ is given, then the *probability that X has a value in the Borel set B* is defined to be $P(X \in B) := P(X^{-1}(B))$. For X and P given, the map $B \mapsto P(X \in B) =: \mu_X(B)$ is a probability measure on $\mathbb{R}$ which is called the *distribution of X (with respect to P)*. The phrase in parentheses is often omitted since P is implicit in many contexts. The probability measure P, being a measure, has a theory of integration that comes with it for free. So, integrals $\mathcal{E}(X) := \int_\Omega X(\omega) \, dP(\omega)$ are defined for a wide class of $X \in \mathcal{M}(\Omega)$, including all bounded, Borel measurable functions. We say that $\mathcal{E}(X)$ is the *expected value* of X (with respect to P).

An identity for the expected value is

$$\mathcal{E}(X) = \int_\Omega X(\omega) \, dP(\omega) = \int_\mathbb{R} \lambda \, d\mu_X(\lambda)$$

in the sense that if one of these two integrals exists, then so does the other and the equality holds. This then expresses the expected value as the first moment of a probability measure on $\mathbb{R}$. More generally, for any bounded, Borel function $f : \mathbb{R} \to \mathbb{C}$ and all $\omega \in \Omega$ we define $f(X)(\omega) := f(X(\omega))$, which is itself bounded and Borel measurable. Then μ_X satisfies

$$\mathcal{E}(f(X)) = \int_\Omega f(X)(\omega)\, dP(\omega) = \int_\mathbb{R} f(\lambda)\, d\mu_X(\lambda).$$

The observables in quantum theory are self-adjoint operators $T = T^*$, but the condition $T \in \mathcal{L}(\mathcal{H})$ is not required though it may hold. The spectrum of any self-adjoint operator (bounded or unbounded) is a non-empty, closed subset of $\mathbb{R}$, and its elements correspond to the values seen in the observations associated to T. For each Borel subset B of $\mathbb{R}$ there is an event in $\mathcal{L}(\mathcal{H})$ that is denoted as $T \in B$ and is called the event that T is observed to have a value in B. As explained in detail below, it is defined by $T \in B := P_T(B)$, where P_T is the pvm associated to T by spectral theory. If a density matrix ρ is given, then the probability that T has a value in the Borel set $B \subset \mathbb{R}$ is defined by Born's Rule to be $P_\rho(T \in B) = Tr(P_T(B)\,\rho)$.

In quantum theory, only rarely would ρ be omitted from this notation on the left side. Probability theory in quantum theory exists even prior to choosing a state ρ, since a self-adjoint operator T has its unique associated pvm P_T. This has properties similar to those of a probability measure, except that it takes values that are quantum events. So, a pvm has a *codomain* consisting of quantum events. On the other hand a classical probability measure has classical events in its *domain*. The integral $\int_\mathbb{R} f(\lambda)\, dP_T(\lambda)$ exists in $\mathcal{L}(\mathcal{H})$ for a wide class of Borel measurable functions $f : \mathbb{R} \to \mathbb{C}$, including all bounded, measurable functions. The 'expected value' of this pvm gives

$$T = \int_\mathbb{R} \lambda\, dP_T(\lambda)$$

by the spectral theorem. Maybe you did not expect this result, which is also called the *diagonalization* of T. Actually,

$$f(T) := \int_\mathbb{R} f(\lambda)\, dP_T(\lambda)$$

defines a *functional calculus* for all bounded, Borel functions $f : \mathbb{R} \to \mathbb{C}$. In the special case for real valued f it turns out that $f(T)$ is self-adjoint. Given that T represents an observable with values in $\operatorname{Spec} T$, its spectrum, then $f(T)$ is understood to represent the observable which applies the function f to the observed value of T. In particular, $\chi_B(T)$, where χ_B is the characteristic function of a Borel subset B of $\mathbb{R}$, represents the observable that T takes a value in B. But we have

$$\chi_B(T) = \int_{\mathbb{R}} \chi_B(\lambda)\, dP_T(\lambda) = \int_B dP_T(\lambda) = P_T(B).$$

So, this justifies the assertion that $P_T(B) = T \in B$ is the observable (and the event) that T takes a value in B as stated above and after Theorem 2.3.1.

Another curious point is that a classical probability measure P satisfies $0 \leq P(A) \leq 1$ for every event A. This is an inequality of real numbers. On the other hand, a pvm P on $\mathbb{R}$ satisfies $0 \leq P(B) \leq I$ for every Borel subset of $\mathbb{R}$. This is an inequality of self-adjoint operators. So, the linearly ordered interval of real numbers $[0,1]$ for probabilities in the classical case is replaced by the lattice of projections in the 'interval' $[0,I]$ of self-adjoint operators. An even more curious point is that the interval $[0,1]$ only contains real numbers, while the 'interval' $[0,I]$ contains self-adjoint operators that are not projections. This opens the door to considering positive operator valued measures, which will not be discussed further.

Yet another way of relating quantum probability to classical probability is by restricting a pvm to a sub-lattice $\mathcal{E}'$ of the lattice events of $\mathcal{E}$ such that $\mathcal{E}'$ is a σ-algebra. Then one can put any classical probability measure whatsoever on $\mathcal{E}'$. This probability measure need not be the restriction of a spectral measure associated to a pvm defined $\mathcal{E}$, in which case one is considering a structure unrelated to quantum theory. However, if one starts with a spectral measure on $\mathcal{E}$, one can restrict it to many such σ-algebra sub-lattices in order to 'view' the pvm in a variety of classical ways. This could be what some would call complementarity, though such a specific description is not usually given. Conversely, one could have a classical probability measure on $\mathcal{E}'$ and ask whether this is the restriction of a spectral measure on $\mathcal{E}$ and, if it is, then whether that spectral measure is unique. This is close to the setting of the Kadison–Singer conjecture (see [23]), which is now a proved theorem (see [29]). In that context one has a *maximal commutative* sub-$*$-algebra $\mathcal{A}$ of a C^*-algebra $\mathcal{C}$ and a state $\phi : \mathcal{A} \to \mathbb{C}$. Then the theorem says that there exists a unique extension $\tilde{\phi} : \mathcal{C} \to \mathbb{C}$ of ϕ which is a state. Colloquially, under these hypotheses one commutative 'snapshot' of a state suffices to characterize it.

Also notice that classical probability measures on $\mathbb{R}$ arise naturally in quantum probability, but classical probability does not involve quantum probability. And finally in classical probability theory, there is no basic time evolution equation, although time dependent stochastic processes are a part of that theory. However, quantum probability is intrinsically a part of quantum theory, which has time evolution as a major facet of the theory. In fact, the Generalized Born's Rule (2.10.2) is the fundamental time evolution equation of quantum theory.

This section may provide the reader with a false intuition about the role of probability in quantum theory. However, the only points of contact with classical probability concern single events, which in themselves are totally irrelevant for the understanding of correlations. In quantum theory we have new definitions for consecutive and conditional probability. The properties of these multi-event probabilities include new, very possibly non-intuitive, features absent in classical probability theory, such as interference terms and dependence on the order of events.

2.16　Expected Value

Expected value is a mostly unremarkable, quite secondary aspect of quantum probability. However, many times I have heard colleagues speak about it incorrectly. Unfortunately, this misunderstanding can be found in print, too. So, I think that it is necessary to give a clarification of this topic.

First, let's repeat what quantum theory says about probability, namely Born's Rule for ψ a unit vector, $A = A^*$ and B a Borel subset of $\mathbb{R}$ which with a bit of new notation is $\mu(B) := P_\psi(A \in B) = \langle \psi, P_A(B)\psi \rangle$. As a function of the Borel subset B of $\mathbb{R}$ with both A and ψ fixed, μ is a probability measure on the real line $\mathbb{R}$. We have seen this basic point already many times. One can now apply concepts from standard probability theory to this probability measure μ. And this is indeed done. For example, one can consider the *moments* of any probability measure. So, for every integer $k \geq 1$ we define the *kth moment* of μ as $m_k := \int_{\mathbb{R}} \lambda^k \, d\mu(\lambda)$, provided that this integral converges absolutely. The *expected value* of μ is defined to be m_1, the first moment, provided again that the integral converges absolutely. The idea is that m_1 is a *statistical estimator* given by probability theory of the empirically observed *sample average*

$$\overline{m} := \frac{\lambda_1 + \cdots + \lambda_n}{n}$$

of $n \geq 1$ measured values (or *sample*) $\lambda_1, \ldots, \lambda_n$. The values measured are typically not all the same and, indeed, this is what motivates one to turn to probability theory in order to understand them. It is quite common that the expected value is not going to be equal to any of the measured values, since this is a common property of the sample average. For example, the sample average of the number of children per family (in a sample of families in a city, say) could be 2.3, which is not an integer.

As a brief aside, let me note that one can continue by defining the *central moments* for every integer $k \geq 2$ by $\sigma_k := \int_{\mathbb{R}} (\lambda - m_1)^k \, d\mu(\lambda)$ provided that the expected value m_1 exists and that this integral also is absolutely convergent.

A frequently used central moment is σ_2, also called the *variance*. And the *standard deviation* of μ is defined by $\sigma := +\sqrt{\sigma_2}$. Standard deviations (or equivalently variances) enter into the Roberts inequality that expresses mathematically the Heisenberg Uncertainty Principle.

It is well known that the moments (or equivalently the central moments) do not uniquely determine μ in all cases. Even in the most favorable case when these moments exist for all k this moment problem might not have a solution μ and, if it does, that solution might not be unique. Colloquially, one can say that the moments carry some information about the probability measure μ, but not in general all the possible information.

We have enough context now for discussing a common misunderstanding. The false assertion is often made that the only information that quantum theory provides about a system is the expected value of observables in a given state. Actually, Born's Rule provides the probability measure for any observable in a given state, which is a lot more than just the first moment of that probability measure. This aspect of quantum theory dates back at least to 1932 in the seminal book [41] of von Neumann and so should be well known in the physics community. My experience is that it is known by some, but not by others. An example of this confusion is the 'proof' that Schrödinger's model is equivalent to the Heisenberg model by showing that the expected value is the same in both models. While this is a necessary condition, by itself it is not sufficient.

To complete this section here is the derivation of the usual formula for calculating the expected value of $A = A^*$ given a pure state ψ in the dense domain of A. We see that

$$m_1 = \int_{\mathbb{R}} \lambda \, \langle \psi, dP_A(\lambda) \, \psi \rangle = \left\langle \psi, \left(\int_{\mathbb{R}} \lambda \, dP_A(\lambda) \right) \psi \right\rangle = \langle \psi, A\psi \rangle.$$

2.17 Dynamics: The Generalized Born's Rule, The Final Version

The *dynamics*, also known as the *time evolution*, of a physical system is given in quantum theory by a further generalization of the time dependent

Generalized Born's Rule (2.10.2). Contrary to the confused opinion of many authors (including me in [39]), there is only one fundamental time evolution equation in quantum theory. It is the same equation in all models. It is the only equation which we subject to experimental verification. But (2.10.2) is still not adequate for all purposes, since it assumes that there is no time evolution of the quantum system between events. So, now we suppose that a system is given with a state represented by ψ at time t_0 and that we are interested in the events $E_1, \ldots, E_n$ at the times $t_1, t_2, \ldots, t_n$, where $t_0 < t_1 < \cdots < t_n$. The probability of this sequence of events at these times is

$$P_\psi(E_1, \ldots, E_n, t_0, t_1, \ldots, t_n)$$
$$= ||E_n U(t_n, t_{n-1}) \cdots E_2 U(t_2, t_1) E_1 U(t_1, t_0)\psi||^2$$

where $U(s, t)$ is the time evolution operator of the system for the times $s < t$. In the Schrödinger model $U(s, t) = \exp(-i(t - s)H)$, where H is the Hamiltonian of the quantum system. I imagine that this formula is obvious to those who think in the Schrödinger model. But whether this formula is obvious or not is an independent consideration. It can be used in this final version of the *Generalized Born's Rule*, which is Wigner's Rule. See [44].

Axiom 5 Updated: Let ρ be a density matrix. Suppose that $E_1, \ldots, E_n$ are events occurring at times $t_1 < \cdots < t_n$. Suppose that $t_0 < t_1$ and that we have *time evolution operators* $U(s, t) \in \mathcal{L}(\mathcal{H})$ for all $s < t$. We define the *consecutive probability with time evolution* to be

$$P_\rho^{t_0, t_1, \ldots, t_n}(E_1, \ldots, E_n) := Tr(E_n U(t_n, t_{n-1}) \cdots E_1 U(t_1, t_0)\rho).$$
$$(2.17.1)$$

The *conditional probability with time evolution* reads much like (2.10.2) with the time evolution operators appropriately interspersed between the event operators. The exact, general formula becomes unwieldy to write down. ∎

This axiom holds in all models if we extend the group E_t to act on all bounded operators and so, in particular, on the operators $U(s, t)$. In the standard models (Schrödinger, Heisenberg, interaction) the action of E_t is via conjugation by operators in a unitary group. This conjugation acts on all of $\mathcal{L}(\mathcal{H})$ as well. As we have come to expect, the value of the probability (2.17.1) is invariant under isomorphisms of models. It might appear strange that the time evolution operators depend on the model, rather than being the same in all models as in the familiar Schrödinger model. But this is exactly one of the things that happens in the interaction model, which also should be familiar for the reader. Recall that in the interaction model one writes the Schrödinger Hamiltonian as $H = H_{free} + H_{int}$, the sum of a 'free' term and of an 'interacting' term. Then the operators in the Schrödinger model, including the time evolution operators, are transformed to operators in the interaction model by conjugating them with $\exp(-itH_{free})$. The splitting of H into two terms is chosen to facilitate subsequent calculations, not to change the results. And that is exactly the point of having different isomorphic models of quantum theory.

Again, the manner for computing this time dependent probability does depend on the model, which leads to much confusion. In the Schrödinger model, which is the most familiar and most widely used model, all the events are time independent and only the state could possibly change with time. As is well known the time evolution of the state in this model is given by a family of equations known collectively as Schrödinger's equation, which all have the same form $i\,\psi'(t) = H\psi(t)$, where H is a self-adjoint operator, known as the Hamiltonian, acting in the Hilbert space. The multitude of physical systems covered by this approach is due to the fact that physicists are very adept at finding the appropriate Hamiltonian for many systems. However, the time dependent state which solves Schrödinger's equation has no physical significance. It is an artifact of the model and nothing else. But it is one of the ingredients that for the case $\rho = |\psi\rangle\langle\psi|$ in (2.17.1) goes into computing the time dependent probability, which indeed does have physical significance. In other models one must calculate the time dependence of the elements in (2.10.2) using other auxiliary equations. But again these auxiliary time dependent elements do not have any physical significance.

It will surely be taken as heresy on my part to say that Schrödinger's equation is without physical significance. However, if that is the fate of its solutions, then that must be the fate of Schrödinger's equation itself. It is a stepping stone, a useful tool. This idea flies in the face of long-standing traditions in physics, especially those that favor differential equations as the most fundamental elements of a physical theory. There is an expectation that the time evolution of a physical system should be expressed as a differential equation involving time as one of the variables. But (2.17.1) does not have that form. Taking the derivative of (2.17.1) in order to find a differential equation that it must solve leads to a relation of the time derivative of the probability with the time evolutions of the events and of the state. But these last two time evolutions are model dependent.

This preference for differential equations manifests itself in the way the equation of motion is written in the Heisenberg model. Typically this is presented as an ill-defined differential equation whose so-called 'solution' is then given. It is that 'solution' (1.7.3) which is the actual time evolution equation, in spite of the fact that it is not a differential equation.

Another apparent obstacle to taking the Generalized Born's Rule (2.17.1) as the fundamental time evolution equation of quantum theory is that this ignores the history of its discovery. The physics community is fascinated with the story of how these ideas emerged and who gets the credit for each of them. And I am complicit in this tradition to the extent that I do use the names of scientists when discussing their discoveries. At one point while writing this treatise I thought of removing 'Lie' from 'Lie group' to give one example of how one might eliminate history from this narrative. Born's Rule in the traditional narrative is seen as a later add-on, an embellishment of ideas that already 'worked' but somehow seemed incomplete. And besides that there already was a known equation, namely Schrödinger's equation, that was recognized by one and all as the time evolution *differential* equation of quantum theory. And this history then continues with Born's Rule being severely criticized by some and rejected by others. This paragraph is one of my few excursions into the history of quantum theory, and the point behind it is that the historical sequence of discoveries is not the logical structure of quantum theory. The Generalized Born's Rule (2.17.1) is the fundamental time evolution equation of quantum theory.

If you think that my purpose here is to remove Schrödinger's equation from its central position in quantum theory, then you are reading correctly. For example, on Wikipedia the various topics on quantum theory (which there is called quantum mechanics) are presented in a box with Schrödinger's equation at its head. This should be replaced with the Generalized Born's Rule according to my thesis.

It is not well understood that one is actually only using the conditional probability (2.10.2) when analyzing many physical phenomena. Let's recall how that works in the case of two events. Given an event E_1 and a unit vector ψ, the conditional probability of a subsequent event E_2 is

$$P_\psi(E_2 \mid E_1) = ||E_2 E_1 \psi||^2 / ||E_1 \psi||^2 = ||E_2 \tilde{\psi}||^2 = P_{\tilde{\psi}}(E_2),$$

where $\tilde{\psi} := E_1 \psi / ||E_1 \psi||$ is the 'collapse' of the original state represented by ψ. Of course, this is just reading backwards one motivating argument for the definition of conditional probability. The point now is that the right side of this equation can always be translated into the conditional probability on the left side. Another common way of speaking of this situation is that the first event E_1 'prepares' a state $\tilde{\psi}$, which the event E_2 then 'measures'. However, it all comes down to quantum conditional probability of quantum events.

Another curious aspect of quantum theory is that the one-event Born's Rule in Axiom 4 is irrelevant for understanding observations. Actually a more careful analysis always reveals that one is considering a conditional probability of two or more events. Science is based on correlations. The statements in this paragraph are meant to be controversial.

2.18 Quantum Information

As with any other aspect of basic quantum theory, information must be defined in a way that is invariant under isomorphisms of models. Since the only structure left invariant is the Generalized Born's Rule, information must be defined in terms of it. Most studies of quantum information are done in the setting of a tensor product of a finite number of finite dimensional Hilbert spaces, where more tools such as partial trace are available. But that restriction is not made here. According to the approach of this treatise, any quantitative measure of information (or of anything else!) calculated using states must be expressible in terms of the Generalized Born's Rule in order to be model independent and therefore physically meaningful. Therefore, any function of quantum probabilities is compatible with this approach. This includes all the usual definitions of entropy.

Any theoretical measure of information not consistent with this approach can and should be checked experimentally in order to adjudicate differences.

2.19 Afterthoughts on Events

Theoretical events correspond to something happening, namely a physical event, in a small region of spacetime. As such the theoretical probabilities of sequences of theoretical events should reflect commonly known facts about the relative frequencies of sequences of physical events. However, theoretical events should not to be confused with other theoretical concepts, such as propositions and operations, which describe other aspects of the physical world. A familiar example is someone getting dressed, who is going to do these two operations:

A: Put on shoes.   B: Put on socks.

This example shows that the two orders in which these two operations can be performed give different results. In fact, by observing someone fully dressed, including socks and shoes, one can tell which order had been used. So, while operations A and B can be thought of as physical events, they are also operations whose order impacts the final 'state' of the dressed person. We might well expect that the order of first A, then B does not occur very often.

But here is another example for someone getting dressed who uses these two operations:

A: Put on shoes.   B: Put on pants.

These can also be thought of as events, but now the resultant 'state' of the dressed person does not indicate the order in which they occurred. But the relative frequency of first A and then B is not necessarily equal to (and we might expect it to be very much less than) the relative frequency of first B and then A. So, the theoretical probabilities assigned to these two physically possible sequences (given some theory of getting dressed!) should reflect this difference and should not be equal. Nonetheless, basic measure theoretical probability theory does not accommodate the order of events, even though subsequent theoretical constructs, such as stochastic processes, do. In that basic theory the probability of these two events is the probability of this single event:

C: Put on shoes and put on pants.

But this is the same as the event

D: Put on pants and put on shoes.

because of a property of the logical connective 'and' that tells us that the *propositions* C and D are logically the same. This commutativity property of 'and' holds in both classical Boolean logic as well as in quantum logic.

However, the quantum consecutive probability as expounded here has the property that the probability of two events can, and often does, depend on the order in which they occur. And this is related to the fact that a sequence of two events in this theory is not itself an event in general.

The examples in this section are meant to be illustrative of our usual ways of thinking about physical events. They are not supposed to be susceptible to analysis as quantum events, though they may be.

2.20 The Fate of the State

> Something is rotten in the state of Denmark.
>
> Marcellus in *Hamlet*
>
> William Shakespeare

The usual understanding is that the state is the complete time dependent description of a quantum system. This idea not only leads to the so-called quantum paradoxes, but it is also in contradiction with the Heisenberg model in which all states are constant in time and so the time evolution given by the Schrödinger equation is irrelevant. This misunderstanding is at the very core of the EPR paper, which only shows that the state (known there as the 'wave function') does not give a complete description of the physical reality of a particular quantum system. Of course not. As with any quantum system, what can happen (events) and how frequently (probabilities) are also a part of its description. The 'wave function' alone does not suffice. (See Chapter 6.)

Even saying that a quantum system is 'in a state' in the Schrödinger model misleads, since this language does not have in general an equivalent in the Heisenberg or interaction models. Unfortunately however, this language is difficult to eliminate given the tendency to find just one deterministic time dependent differential equation whose solution gives a complete description of a physical system. But then this tendency–one might even call it the classical tendency–denies the equivalence of the Schrödinger, Heisenberg and interaction models. So, those who wish to save the Schrödinger model and its language of changing states are logically constrained to reject its equivalence to the Heisenberg and interaction models. This then requires an axiomatic approach in contradiction to that given here such as that found in [16]. The differences between these axiomatizations must be checked by experiment.

States are mathematical objects used to compute probabilities and, most importantly, conditional and consecutive probabilities of multiple events.

2.21   Events Suffice for All Observables

The mathematics on this issue is clear. Self-adjoint operators, which are taken to represent physical observables in standard quantum theory, are in bijective correspondence with pvm's (Definition 2.3.1). The projections in a pvm are (quantum) events. The events in the pvm's are what enter the formulas for probability. So, events (which are theoretical observables) suffice for doing quantum theory. Of course, events include all measurements. Here is an example of two ways to say the same thing. (1) The measured values of a physical observable are the elements in the spectrum of its associated self-adjoint operator. (2) The measured values of a physical observable are the elements in the support of its associated pvm. (See Theorem 2.3.2.)

2.22   The Irrelevance of Single Events

As mentioned earlier, the idea is that science is based on understanding the correlations of multiple events, and so the probability of isolated events is irrelevant. But unfortunately, quantum theory has been dominated by a language that only speaks of isolated events (typically measurements) and their probabilities. A main thesis of this treatise is that only by including both consecutive and conditional probabilities into quantum theory can the inherent confusion induced by the standard dominant language begin to be dispelled. Of course, at times I use the probability of a single event in a formula. But that is always in a context with the probabilities involving other events. This brief section is also meant to stimulate further consideration and possible refinements of this admittedly controversial idea.

2.23   Preparation of the State

The preparation of a state as a way to start an experiment is standard topic in most quantum texts. This seems to be a way of 'choosing' first a pure state, say represented by ψ, in order to calculate the probability $P_\psi(E)$ of a single subsequent event E. This seems to contradict the previous section and so deserves some further comment. But we can see this in terms of the conditional probability of first the event $|\psi\rangle\langle\psi|$ occurring and then E subsequently occurring, given a pure state represented by a unit vector ϕ, as follows:

$$P_\phi(E \,|\, |\psi\rangle\langle\psi|) = \frac{P_\phi(|\psi\rangle\langle\psi|, E)}{P_\phi(|\psi\rangle\langle\psi|)}$$

$$= \frac{||E|\psi\rangle\langle\psi|\phi||^2}{\langle\phi, |\psi\rangle\langle\psi|\phi\rangle}$$

$$= \frac{|\langle\psi,\phi\rangle|^2 ||E\psi||^2}{|\langle\psi,\phi\rangle|^2}$$

$$= ||E\psi||^2 = P_\psi(E)$$

provided that the unit vector ϕ satisfies $\langle\psi,\phi\rangle \neq 0$, that is $\phi \notin \{\psi\}^\perp$. We get a similar result if we use a density matrix ρ instead of ϕ, as follows:

$$P_\rho(E \,|\, |\psi\rangle\langle\psi|) = \frac{P_\rho(|\psi\rangle\langle\psi|, E)}{P_\rho(|\psi\rangle\langle\psi|)}$$

$$= \frac{Tr(|\psi\rangle\langle\psi|E|\psi\rangle\langle\psi|\rho)}{Tr(|\psi\rangle\langle\psi|\rho)}$$

$$= \frac{\langle\psi, E\psi\rangle\langle\psi, \rho\psi\rangle}{\langle\psi, \rho\psi\rangle}$$

$$= \langle\psi, E\psi\rangle = P_\psi(E)$$

provided ρ satisfies $\langle\psi, \rho\psi\rangle \neq 0$. This shows how the single event probability $P_\psi(E)$ of an event E given a 'prepared state' ψ can be understood as the conditional probability of the sequence $|\psi\rangle\langle\psi|$ and then E. This depends on the fortuitous mathematical fact that pure states are in bijection with rank one density matrices, which are events. So, 'the state represented by ψ was prepared' can be translated into 'the event $|\psi\rangle\langle\psi|$ occurred'.

However, the analysis of the previous paragraph is only applicable if the event $|\psi\rangle\langle\psi|$ is an element in the von Neumann algebra of the system. It seems this is the case in the uses of the phrase 'preparation of the state' in the literature. But that literature is enormous and so this may be a shortcoming of the present approach. It is also noteworthy that this approach does not accommodate a translation of 'the density matrix was prepared' into a statement about a single experiment.

2.24   Is Standard Quantum Theory Being Changed?

No, it certainly is not. All of the calculations and predictions of probabilities given by standard quantum theory remain the same, provided that one uses the Schrödinger model or any model isomorphic to it. In the Schrödinger model one has available the standard mathematical methods just as before, namely Schrödinger's equation and the collapse of the state. What is different is that it would be better to speak about these methods in a new way for what they are: mathematical methods and not physical processes.

In my introductory text [39] I used the expression *Schrödinger's razor* to describe the desire by Schrödinger to find a partial differential equation which would describe quantum phenomena just via its solutions without any extra, auxiliary conditions. For a beginning student I think it is appropriate to present Schrödinger's equation as that differential equation. In the Hilbert space context there is then an elegant existence and uniqueness theorem of its solutions for self-adjoint Hamiltonians. And this is rather gratifying both physically and mathematically. But extra rules must be given, namely Born's Rule in order to compute one-event probabilities and collapse of the state in order to compute conditional probabilities. While this spoils Schrödinger's original intention, it is essential in making the Schrödinger model work and in also making it isomorphic to the Heisenberg model. It turns out Schrödinger's equation comes very close to fulfilling the desire of having a fundamental quantum time evolution equation that is also a differential equation. The utility of Schrödinger's equation as a practical tool remains unchanged. Still, the fundamental time evolution equation of quantum theory, the Generalized Born's Rule, is not a differential equation. And that is not a bad thing.

While I might be criticized for throwing differential equations to the wolves, I wish to emphasize their important, continuing role in science and, in particular, in physics. But we all have become so accustomed to them that we, just like Schrödinger, find it difficult to imagine a viable alternative.

Yet, Maxwell's equations are instructive in this regard. These are usually presented as differential equations, although I first learned about them as integral equations. For me the transition to the differential formulation was neither intuitive nor motivated. Besides those who have to teach what the differential Maxwell equations 'mean' often translate them into their integral form as a pedagogical aid. After all, how else can a student understand what $\nabla \cdot \mathbf{B} = 0$ is saying? Or $\nabla \cdot \mathbf{E} = 4\pi\rho$? Or what their difference is all about?

In conclusion I would prefer to avoid speaking at all of the collapse of the state and instead to replace that language with references to quantum conditional probability. However, there is no harm done if one wishes to continue to speak of collapse as long as one realizes that this is nothing more than code language for a mathematical operation used as a step in computing a quantum conditional probability.

One should be cognizant that a mathematical concept need not have any physical meaning or 'interpretation'. As an example in the context of Hilbert space theory, the equation[3] $\psi + \psi = 2\,\psi$ makes mathematical sense and is correct. Yet, as far as I am aware, it has no meaning in quantum theory. Another basic operation in Hilbert space theory is the inner product. Again, this has no physical 'interpretation'. The collapse of the state and, sadly, Schrödinger's equation do not have either one in and of themselves any physical meaning. However, by combining them they do allow us to calculate time dependent probabilities using the Generalized Born's Rule, which does have a physical meaning. And in this sense standard quantum theory remains unchanged. Explicitly, the philosophy of 'shut up and calculate' is vindicated since all the results of standard quantum mechanics remain valid in this approach.

---

[3]This is after all just $1 + 1 = 2$ in quantum clothing.

Chapter 3

Entanglement

> I would not call that property[1] **one** but rather
> **the** characteristic trait of quantum mechanics,
> the one that enforces its entire departure from
> classical lines of thought. (emphasis in original)
>
> — Erwin Schrödinger

Entanglement has been one of the most puzzling topics in quantum theory. This is largely due to a detailed discussion about two events that presumably 'know' nothing of each other but nonetheless show an uncanny deterministic relation between them. One wonders how a probabilistic theory could ever explain this 'determinism'. Worse yet, these two events can be space-like with respect to each other and so relativity theory forbids information from passing from one to the other. Yet, in some strange form, this seems to be what is happening. There are many experiments that agree with entanglement, including some carefully constructed experiments where the two events are indeed space-like. And no experiment, as far as I am aware, contradicts entanglement. How can we deal with this conundrum? Must quantum theory be modified or abandoned? Is there some spooky action at a distance that accounts for entanglement? Is physics non-local? How can the collapse of a state by one detector possibly affect another detector?

Entanglement can be understood using just basic quantum theory: events, states and probability. The point is that in any probability theory, some of the calculated probabilities will turn out to be 1. Experimental confirmation of entanglement should not produce existential *angst*, much less to appeals for new scientific principles to 'explain' the 'mystery' of this

---

[1]The property referred to is entanglement. See Ref. [37].

confirmation. So, what is going on here in terms of basic quantum theory? Quite simply put, entanglement always involves *three* events, never just two. (Or possibly more than three!) There always is an event that precedes in time the two other events. Of course, those two later events 'know' about that earlier event. When the calculated conditional quantum probability of these three (or more) events shows that they are not independent, this lack of independence is defined as *entanglement* of these events. And this agrees with the experiment.

3.1 A Standard Example

Here is some notation for a standard example of entanglement. We have an event E_0, which we will call the initial event, and then two events E_1 and E_2, each of which occurs after E_0. The time relation between E_1 and E_2 is irrelevant to this analysis. This example captures the essence of the EPR paper and is discussed further in Chapter 6. We could assume that E_1 and E_2 commute, but that assumption is not used until the end of this chapter. Note that space-like separated events *must* commute according to special relativity, while *some* time-like (or light-like) separated events *can* commute.

We start by taking $\mathcal{H} = \mathbb{C}^2 \otimes \mathbb{C}^2$, which is the Hilbert space for the quantum system. ($\mathbb{C}^2$ and $\mathcal{H}$ have the standard inner products.) We let S_1, S_2 and S_3 denote the standard 2×2 spin matrices. These self-adjoint matrices are taken as acting on $\mathbb{C}^2$, which is not the Hilbert space $\mathcal{H}$ of this quantum system, and so they are not observables. In the sequel, we only use

$$S_3 = \frac{1}{2}\begin{pmatrix} 1 & 0 \\ 0 & -1 \end{pmatrix}$$

for convenience in doing calculations. Similar results hold for the other spin matrices. Then, we next consider these self-adjoint operators acting on $\mathcal{H}$, which are observables of this quantum system:

$$S^2, \qquad S_r := S_3 \otimes I_2, \qquad S_l := I_2 \otimes S_3,$$

where I_2 is the identity operator acting on $\mathbb{C}^2$ and where the Casimir operator

$$S^2 = (S_1 \otimes S_1)^2 + (S_2 \otimes S_2)^2 + (S_3 \otimes S_3)^2$$

is another standard self-adjoint spin operator. The three events acting in $\mathcal{H}$ which we will consider are defined as

$$E_0 := P_{S^2}(\{0\}), \qquad E_1 := P_{S_r}(\{1/2\}), \qquad E_2 := P_{S_l}(\{-1/2\}).$$

The way one verbalizes this can be misleading, but I will venture to do it anyway. The event E_0 says that initially, the system has total spin 0. The event E_1 says that a detector to the right of the place where E_0 occurred measures the z component of the spin to be $1/2$. The event E_2 says that a detector to the left of the place where E_0 occurred measures the z component of the spin to be $-1/2$. The words 'right' and 'left' merely give a visual rendition to the formalism; they can be replaced with 'Alice' and 'Bob'.

Note that in this example the self-adjoint operators S_r and S_l commute. Consequently, the events E_1 and E_2 also commute.

Given these definitions, it is an exercise, which we now do, to compute the appropriate conditional probability for these three events. The reader should be aware that the intermediate steps in this calculation are without physical significance. The only physically significant part of this calculation is the last step where the value of the conditional probability is given.

Let $\varepsilon_1, \varepsilon_2$ be the standard orthonormal basis of $\mathbb{C}^2$, namely

$$\varepsilon_1 = \begin{pmatrix} 1 \\ 0 \end{pmatrix} \quad \text{and} \quad \varepsilon_2 = \begin{pmatrix} 0 \\ 1 \end{pmatrix}.$$

These unit vectors do *not* represent states in the quantum Hilbert space $\mathcal{H}$. We first identify the pertinent quantum events in terms of these vectors using Dirac notation:

$$E_1 = P_{S_r}(1/2) = |\varepsilon_1\rangle\langle\varepsilon_1| \otimes I_2 \quad \text{and} \quad E_2 = P_{S_l}(1/2) = I_2 \otimes |\varepsilon_2\rangle\langle\varepsilon_2|.$$

Note that these are rank 2 projections. The operator S^2 is not central in $\mathcal{L}(\mathcal{H})$. It is well known that its spectrum is $\{0, 2\}$. The eigenspace for the eigenvalue 2 has dimension 3 (a triplet for spin 1), while the eigenspace for the eigenvalue 0 has dimension 1 (a singlet for spin 0), and this latter eigenspace has an orthonormal basis consisting of this one unit vector:

$$\psi_0 := \frac{1}{\sqrt{2}}(\varepsilon_1 \otimes \varepsilon_2 - \varepsilon_2 \otimes \varepsilon_1). \tag{3.1.1}$$

So, we have the quantum event

$$E_0 = P_{S^2}(\{0\}) = |\psi_0\rangle\langle\psi_0|,$$

which is a projection in $\mathcal{H}$ with rank 1. We next compute the quantum conditional probability

$$P_\psi(E_1 \mid E_2, E_0) = \frac{||E_1 E_2 E_0 \, \psi||^2}{||E_2 E_0 \, \psi||^2} \qquad (3.1.2)$$

provided that $||E_2 E_0 \psi|| \neq 0$. (When not stated otherwise, $\psi \in \mathcal{H}$ is a unit vector.) Intuitively, the expression (3.1.2) gives the conditional probability that the detector on the right measures the z-component of spin to be $1/2$ given that the detector on the left measures the z-component of spin to be $-1/2$ and that (previously) the initial state has been prepared to have spin 0. These words in everyday language only serve to reassure the dubious that this formalism has its motivation. We first have for all $\psi \in \mathcal{H}$ that $E_0 \psi = |\psi_0\rangle\langle\psi_0 |\psi = \langle\psi_0, \psi\rangle\psi_0$. Continuing we have for all $\psi \in \mathcal{H}$ that

$$\begin{aligned}
E_2 E_0 \psi &= (I_2 \otimes |\varepsilon_2\rangle\langle\varepsilon_2|)\langle\psi_0, \psi\rangle \, \psi_0 \\
&= 2^{-1/2}\langle\psi_0, \psi\rangle(I_2 \otimes |\varepsilon_2\rangle\langle\varepsilon_2|)(\varepsilon_1 \otimes \varepsilon_2 - \varepsilon_2 \otimes \varepsilon_1) \\
&= 2^{-1/2}\langle\psi_0, \psi\rangle(\varepsilon_1 \otimes |\varepsilon_2\rangle\langle\varepsilon_2|\varepsilon_2 - \varepsilon_2 \otimes |\varepsilon_2\rangle\langle\varepsilon_2|\varepsilon_1) \\
&= 2^{-1/2}\langle\psi_0, \psi\rangle(\varepsilon_1 \otimes \varepsilon_2) \\
&= 2^{-1/2}|\varepsilon_1 \otimes \varepsilon_2\rangle\langle\psi_0|\psi.
\end{aligned}$$

This is non-zero if and only if $\langle\psi_0, \psi\rangle \neq 0$, which we assume from now on. We conclude

$$E_2 E_0 = 2^{-1/2}|\varepsilon_1 \otimes \varepsilon_2\rangle\langle\psi_0|. \qquad (3.1.3)$$

Clearly, $E_2 E_0$ is a non-zero, rank 1 operator that is not a projection. Finally, for all $\psi \in \mathcal{H}$, we further evaluate that

$$\begin{aligned}
E_1 E_2 E_0 \psi &= (|\varepsilon_1\rangle\langle\varepsilon_1| \otimes I_2)2^{-1/2}\langle\psi_0, \psi\rangle(\varepsilon_1 \otimes \varepsilon_2) \\
&= 2^{-1/2}\langle\psi_0, \psi\rangle(|\varepsilon_1\rangle\langle\varepsilon_1| \otimes I_2)(\varepsilon_1 \otimes \varepsilon_2) = 2^{-1/2}\langle\psi_0, \psi\rangle(\varepsilon_1 \otimes \varepsilon_2) \\
&= 2^{-1/2}|\varepsilon_1 \otimes \varepsilon_2\rangle\langle\psi_0|\psi = E_2 E_0 \psi.
\end{aligned}$$

This implies that we have two equal *operators* (not projections)

$$E_1 E_2 E_0 = E_2 E_0. \qquad (3.1.4)$$

This is more than we need to conclude that the conditional probability (3.1.2) is equal to 1, namely

$$P_\psi(E_1 \mid E_2, E_0) = \frac{||E_1 E_2 E_0 \, \psi||^2}{||E_2 E_0 \, \psi||^2} = 1.$$

The equality (3.1.4) of these two operators is an intermediate step which has no physical significance. For example, the operator $E_2 E_0$ is *not* an event since E_0 and E_2 do not commute. Note that E_0 and E_1 also do not commute. This lack of commutativity is good news. It means that we are dealing with a truly quantum situation.

Similarly, one can calculate that

$$P_\psi(E_2 \mid E_1, E_0) = \frac{||E_2 E_1 E_0 \, \psi||^2}{||E_1 E_0 \, \psi||^2} = 1$$

provided that $||E_1 E_0 \, \psi|| \neq 0$. The details are left to the reader. Using this straightforward method for 3 events with the same time order, one can evaluate the conditional probability for other combinations of components of spin. Typically, those probabilities will lie strictly between 0 and 1, but in the specific example given here, a value of 1 is the result. The fact that a certain probability is 1 does not invalidate the use of probability theory but is merely a special case of it. In particular, probability 1 does not imply that this is a deterministic situation in the sense that the past determines the present. Here, the event E_2 is not necessarily in the past of the event E_1 nor *vice versa*. However, the event E_0 is in the past of both E_1 and E_2.

In other specific cases, probability 0 will occur. For example, we have

$$P_\psi(E_1^c \mid E_2, E_0) = \frac{||(I - E_1) E_2 E_0 \, \psi||^2}{||E_2 E_0 \, \psi||^2} = 0,$$

where $E_1^c = I - E_1$ is the event complementary to E_1. One of the intermediate steps used to show this (using a formula above) is $(I - E_1) E_2 E_0 = 0$.

In any probabilistic theory, some of the calculated probabilities will turn out to be 1. I do not see anything 'spooky' about that. Nor do I see anything 'non-local' in this analysis. Nor do I see that an element of 'reality' is required to 'explain' a probability 1 situation. If you are seeking 'reality' here, why not say that *all* probabilities are real? See Chapter 6 for more on this.

Of course, we can discard the information about the event E_2 and ask only for the conditional probability of E_1 given E_0. Intuitively, this is the point of view of the experimenter on the right. So, we want to compute

$$P_\psi(E_1 \mid E_0) = \frac{||E_1 E_0 \, \psi||^2}{||E_0 \, \psi||^2}.$$

We first note that we have $E_0 \psi = \langle \psi_0, \psi \rangle \psi_0$ for all $\psi \in \mathcal{H}$ as before. Next, we calculate for all $\psi \in \mathcal{H}$ that

$$
\begin{aligned}
E_1 E_0 \psi &= \big(|\varepsilon_1\rangle\langle\varepsilon_1| \otimes I_2 \big)\big(\langle\psi_0, \psi\rangle \psi_0 \big) \\
&= 2^{-1/2} \langle\psi_0, \psi\rangle \big(|\varepsilon_1\rangle\langle\varepsilon_1| \otimes I_2 \big)\big(\varepsilon_1 \otimes \varepsilon_2 - \varepsilon_2 \otimes \varepsilon_1 \big) \\
&= 2^{-1/2} \langle\psi_0, \psi\rangle \big(\varepsilon_1 \otimes \varepsilon_2 \big) \\
&= 2^{-1/2} |\varepsilon_1 \otimes \varepsilon_2\rangle\langle\psi_0|\psi.
\end{aligned}
$$

Consequently, $E_1 E_0 = 2^{-1/2} |\varepsilon_1 \otimes \varepsilon_2\rangle\langle\psi_0|$ and therefore

$$
P_\psi(E_1 \mid E_0) = \frac{||E_1 E_0\,\psi||^2}{||E_0\,\psi||^2} = \frac{||2^{-1/2}|\varepsilon_1 \otimes \varepsilon_2\rangle\langle\psi_0|\psi\,||^2}{||\langle\psi_0, \psi\rangle\,\psi_0||^2} = 1/2
$$

since $||\,|\varepsilon_1 \otimes \varepsilon_2\rangle\langle\psi_0|\psi\,|| = |\langle\psi_0, \psi\rangle|$. In other words, the axiom for calculating the conditional probability gives the expected result. Similarly, one can show that $P_\psi(E_2 \mid E_0) = 1/2$.

We can also compute the conditional probability that E_1 and E_2 occur given E_0. This conditional probability does not depend on the time order of the events E_1 and E_2 since they commute. The conditional probability for this situation is

$$
P_\psi(E_1, E_2 \mid E_0) = \frac{||E_1 E_2 E_0\,\psi||^2}{||E_0\,\psi||^2}
$$

provided $||E_0\,\psi|| \neq 0$. Recall (3.1.3) and (3.1.4):

$$
E_1 E_2 E_0 = 2^{-1/2} |\varepsilon_1 \otimes \varepsilon_2\rangle\langle\psi_0|.
$$

This equality of *operators*, not of events, has no physical significance but does allow us to proceed to the next part of the calculation. So, here it is

$$
P_\psi(E_1, E_2 \mid E_0) = \frac{||E_1 E_2 E_0\,\psi||^2}{||E_0\,\psi||^2} = \frac{||\,2^{-1/2}|\varepsilon_1 \otimes \varepsilon_2\rangle\langle\psi_0|\psi\,||^2}{||\langle\psi_0, \psi\rangle\,\psi_0||^2} = 1/2.
$$

So, we have calculated a different conditional probability, and we get a probability that is neither 0 nor 1. Note that this result agrees with what little intuition we can muster concerning quantum theory. In possibly misleading ordinary language, it says that after measuring the system to have spin 0, the probability is 1/2 that the detector on the right measures the z-component of spin to be $+1/2$ and the detector on the left measures the z-component of spin to be $-1/2$.

What about the rest of the probability? Simple, similar calculations which are left to the reader show that $P_\psi(E_1^c, E_2^c \mid E_0) = 1/2$, where the *complementary event* of an event E is defined as $E^c := I - E$. One also easily verifies that $P_\psi(E_1^c, E_2 \mid E_0) = 0$ and $P_\psi(E_1, E_2^c \mid E_0) = 0$. All of these probabilistic statements can be expressed in ordinary language, and they all agree with commonly held 'intuition'.

One mathematical fact behind this example is that both events E_1 and E_2 are products of the form $F_1 \otimes F_2$, where each factor is an event in $\mathcal{L}(\mathbb{C}^2)$, but the event E_0 is not of that form. Here, the factorizations are with respect to the given definition $\mathcal{H} = \mathbb{C}^2 \otimes \mathbb{C}^2$. This fact is not invariant in general under a unitary transformation to another isomorphic Hilbert space with its own given factorization as a product of two Hilbert spaces of dimension 2. Nonetheless, it does make sense to speak of entangled events. It does not make sense to speak of entangled particles. A good rule of thumb is that when someone speaks of *two* entangled particles, it is best to rephrase that as a clearer statement about *three* entangled events. Entanglement is a property of the probability of events. Recall that entanglement has been defined in Definition 2.8.3.

Next, I want to rephrase this example in terms of probabilities with no state. Using (2.11.2), the fact that $P(E_0) = 1$ and previously calculated probabilities, we see that

$$P(E_1 \mid E_0) = P_{\psi_0}(E_1) = 1/2,$$

$$P(E_2 \mid E_0) = P_{\psi_0}(E_2) = 1/2,$$

$$P(E_1 E_2 \mid E_0) = P(E_2 E_1 \mid E_0) = P_{\psi_0}(E_1 E_2) = 1/2.$$

This implies that the two *commuting* events E_1 and E_2 are not independent with respect to the earlier event E_0 or, equivalently, with respect to the 'initial' state ψ_0 since

$$P_{\psi_0}(E_1 E_2) \neq P_{\psi_0}(E_1) P_{\psi_0}(E_2).$$

One might be misled into thinking that commuting events should always be independent, but that need not be so since that depends on other prior circumstances. This motivates the following definition.

Definition 3.1.1. Given three events E_0, E_1, E_2 with $E_0 \neq 0$ being earlier than the others, we say that these events are *entangled* if the later two (but not necessarily commuting) events E_1, E_2 are not independent in some order when their probabilities are conditioned on the first event E_0. Equivalently,

either $\quad \Delta(E_1, E_2 \mid E_0) := P(E_1 E_2 \mid E_0) - P(E_1 \mid E_0)P(E_2 \mid E_0) \neq 0$

or $\quad \Delta(E_2, E_1 \mid E_0) = P(E_2 E_1 \mid E_0) - P(E_1 \mid E_0)P(E_2 \mid E_0) \neq 0.$

$$(3.1.5)$$

This is called *entanglement (without any state)*, another new definition.

This definition can be extended to sequences of three or more events by following the idea in Section 2.10. The numbers $\Delta(E_1, E_2 \,|\, E_0)$ and $\Delta(E_2, E_1 \,|\, E_0)$ in the interval $[-1, 1]$ are quantitative measures of 'how much' the events are entangled. According to this new definition, entanglement is nothing other than a probabilistic property of quantum events, namely the lack of independence among them without any reference to a state.

Note that in this definition, the Hilbert space need not be represented as a tensor product. It might be true (possibly plus some reasonable hypotheses) that three entangled events are such that in some unitarily equivalent tensor product Hilbert space, the two later events are factorizable as a tensor product of events, while the earliest one of them is not. However, even though a tensor product representation can be quite useful, it tends to hide the underlying probabilistic character of entanglement.

In this example, the first event E_0 could have rank 1, and so it seems that this event only serves to 'prepare' the 'real' initial state $\psi_0 = E_0 \psi / \|E_0 \psi\|$, where $\psi \in \mathcal{H}$ is some 'pre-initial' state satisfying $E_0 \psi \neq 0$. But thinking of it that way misses the point. States do not enter into this analysis at all. Everything follows from only the conditional probabilities of events. If one wishes to preserve this terminology, one could say an initial event E_0 of any rank has been 'prepared'. But this is just language describing exact formulas.

3.2 Entangled States

> Sir, I have found you an argument
> but I am not obliged to find you
> an understanding.
>
> Samuel Johnson

Of course, anyone familiar with entanglement realizes immediately that ψ_0 in (3.1.1) is an *entangled state*. How does that terminology fit in with the analysis in terms of the conditional probabilities of three events? First of all, to say that ψ_0 in (3.1.1) is an entangled state means by the standard definition that it can not be written as a factorized vector, namely as $\phi_1 \otimes \phi_2$ where ϕ_1 and ϕ_2 are unit vectors in $\mathbb{C}^2$. (This is a standard exercise.) So, the vector ψ_0 is an entangled state not in and of itself but only relative to a given tensor product representation of the Hilbert space of the system.

So, let's see how such a tensor product Hilbert space $\mathcal{H} = \mathcal{H}_1 \otimes \mathcal{H}_2$ gives a special case of entanglement. As usual, we avoid trivial cases by requiring that $\dim \mathcal{H}_j \geq 2$ for $j = 1, 2$. We consider any unit vector $\psi_0 \in \mathcal{H}$, whether it is entangled or not, and from it construct the non-zero event $E_0 := |\psi_0\rangle\langle\psi_0|$. Next, we use the tensor product structure to define events $E_1 := F_1 \otimes I_2$ and $E_2 := I_1 \otimes F_2$, where for $j = 1, 2$, we let $F_j \neq 0$ be any event in $\mathcal{L}(\mathcal{H}_j)$. Also, I_j denotes the identity operator acting on $\mathcal{H}_j$. We require E_0 to be the earliest event. The relative time order of E_1 and E_2 is irrelevant. Clearly, E_1 and E_2 are commuting non-zero events with dimension ≥ 2. In particular, neither E_1 nor E_2 corresponds to a state, even though F_1 or F_2 could correspond to a pure state in its respective Hilbert space. Using (2.11.2) and $P(E_0) = 1$, we have for any E (which is an event or a product of events) that

$$P(E \mid E_0) = P(E\, E_0) = P(E\, |\psi_0\rangle\langle\psi_0|) = P_{\psi_0}(E).$$

Consequently, the conditions in (3.1.5) become respectively

$$\text{either} \quad \Delta_{\psi_0}(E_1, E_2) := P_{\psi_0}(E_1 E_2) - P_{\psi_0}(E_1)P_{\psi_0}(E_2) \neq 0$$

$$\text{or} \quad \Delta_{\psi_0}(E_2, E_1) = P_{\psi_0}(E_2 E_1) - P_{\psi_0}(E_1)P_{\psi_0}(E_2) \neq 0.$$

If one of these two conditions holds, we say that E_1, E_2 are *entangled* with respect to the *initial state* ψ_0. With this notation, we have the next result.

Theorem 3.2.1. *The following statements are equivalent:*

1. ψ_0 *is an entangled state.*
2. *For some choice of events F_1 and F_2, the two events E_1 and E_2 are entangled with respect to ψ_0.*
3. *For some choice of events F_1 and F_2, the three events E_0, E_1, E_2 are entangled.*

Proof. The second and third statements are equivalent by the remarks above. We continue with other cases:

Not 1 $\implies$ Not 2: This was Example 2.8.1.

1 $\implies$ 2: Suppose that ψ_0 is entangled. The Schmidt decomposition applies to Hilbert spaces of all dimensions, whether finite or infinite. (See Ref. [7].) This gives us a non-unique, norm convergent representation of ψ_0:

$$\psi_0 = \sum_{j=1}^{n} \lambda_j\, \phi_j \otimes \psi_j \in \mathcal{H}_1 \otimes \mathcal{H}_2,$$

where for each j we have $\lambda_j > 0$ and ϕ_j (resp., ψ_j) is an orthonormal *set*, but not necessarily a *basis*, of $\mathcal{H}_1$ (resp., $\mathcal{H}_2$). The λ_j's need not be distinct. However, the upper index $n \in \mathbb{N}^+ \cup \{+\infty\}$ is unique. Also, ψ_0 is entangled by definition if and only if $n \geq 2$. We define $F_1 := |\phi_1\rangle\langle\phi_1|$ and $F_2 := |\psi_2\rangle\langle\psi_2|$ and then calculate that

$$P_{\psi_0}(E_1) = ||E_1\psi_0||^2 = ||(|\phi_1\rangle\langle\phi_1| \otimes I_2)|\psi_0||^2 = ||\sum_{j=1}^{n} \lambda_j (|\phi_1\rangle\langle\phi_1|\phi_j\rangle) \otimes \psi_j||^2$$

$$= ||\sum_{j=1}^{n} \lambda_j \delta_{1,j}\phi_1 \otimes \psi_j||^2 = ||\lambda_1(\phi_1 \otimes \psi_1)||^2 = \lambda_1^2 \neq 0,$$

where $\delta_{i,j}$ denotes the Kronecker delta. Similarly, we have

$$P_{\psi_0}(E_2) = \lambda_2^2 \neq 0.$$

And we next see that

$$P_{\psi_0}(E_1E_2) = ||(|\phi_1\rangle\langle\phi_1| \otimes I_2)(I_1 \otimes |\psi_2\rangle\langle\psi_2|)\psi_0||^2$$

$$= ||(|\phi_1\rangle\langle\phi_1| \otimes |\psi_2\rangle\langle\psi_2|)\psi_0||^2 = ||\sum_{j=1}^{n} (|\phi_1\rangle\langle\phi_1| \otimes |\psi_2\rangle\langle\psi_2|)\phi_j \otimes \psi_j||^2$$

$$= ||\sum_{j=1}^{n} |\phi_1\rangle\langle\phi_1|\phi_j\rangle \otimes |\psi_2\rangle\langle\psi_2|\psi_j\rangle||^2 = ||\sum_{j=1}^{n} \delta_{1,j}\phi_1 \otimes \delta_{2,j}\psi_2||^2 = 0 \neq \lambda_1^2\lambda_2^2.$$

Therefore, for these choices of F_1 and F_2, we have that E_1 and E_2 are entangled with respect to ψ_0.   □

The point of this theorem is that this standard entanglement situation *can* be understood in terms of nothing other than probabilities and events, but I do not mean to say that one *must* do it this way. As in all other aspects, this treatise does not discard anything in standard quantum theory. Rather what I am simply trying to do is provide a language that is both inclusive and logical. The mathematics of the entanglement properties of quantum systems gets amazingly complicated. For example, see Refs. [4, 31]. There is no reason to stop using standard techniques for dealing with that. For example, partial trace is an available mathematical tool, which can enter the theory in Axiom 3. But I conjecture that all studies of entangled states can be understood using only events and probabilities since the entangled state itself is simply an initial event which conditions the probabilities of subsequent events. If that turns out to be incorrect, then substantive changes must be made to the axioms proposed in Chapter 1. And that would be quite interesting since such changes must conform with the experiment.

3.3 Collapse Is Marginalized

> Reason itself does not work instinctively, but requires
> trial, practice, and instruction in order gradually
> to progress from one level of insight to another.
>
> Immanuel Kant

The analysis of entanglement given in this chapter does not depend in any way on Schrödinger's equation. Both the states and the events have been taken to be time-independent. So, this example is in both the Schrödinger model and the Heisenberg model. Typically, in other presentations, the time dependence is imposed on the spatial structure rather than on the spin structure in a way consistent with the presentation here. Most importantly, we learn that Schrödinger's equation is not a basic structure in quantum theory, while quantum probability theory is basic.

The critical reader may object that I have done nothing other than just repeat the standard analysis for an entanglement experiment in a disguised form. But that ignores certain key aspects of this presentation. First, it is an analysis of three events, not of two particles. And second, there is no reference to collapse, which is only a way of describing some mathematical intermediate steps in other, more common ways of dealing with entanglement.

And collapse does not require an 'explanation' any more than any other mathematical algorithm that gives the right results. The terminology for this particular step is unfortunate since it leads one to think that some sort of 'understanding' of a physical process is needed. Calculations are meant to help us understand physical phenomena, or as it has been said since Classical Antiquity to save the phenomena. (See Ref. [35].)

But for those who cling to the collapse language, let me note that collapse also occurs in the context of classical physics despite the often made claim that entanglement is a purely quantum effect. One can easily produce two space-like events which are highly correlated and are described classically. As in the quantum case, one simply allows an event in the common past of these two events to have an impact on them. As in the quantum case, the two later events would be binary: one of two related possibilities. The highly touted determinism of classical mechanics has nothing to do with this. I will come back to this later in Chapter 7.

Also, I have done an analysis here in terms of the conditional probability of events, which is never done when analyzing entanglement as far as I am aware. Rather the standard discussion uses these words: particle, detector, measurement and collapse, none of which is needed in this approach. It turns out that the conditional probability in the particular case (3.1.2) just happens to be 1. Again, I want to emphasize strongly that the intermediate steps in the calculation have absolutely no physical significance. Only the calculated probability matters. The rules of quantum probability say that this particular combination of events occurs in 100% of repetitions of the experiment. This deserves to be checked by experiment. And it has been. The conclusion is that quantum theory is verified. This seems to be widely accepted in the physics community.

Let me make clear that entanglement is a property of events and not of anything else. It makes no sense to entangle particles because particles are not events. It makes no sense to entangle the states (note the plural!) of two particles because referring back to Axiom 2 only the entire quantum system itself has *one* associated state and not its constituent particles. It makes no sense to entangle a system with its environment since neither a system nor its environment is an event. It makes no sense to entangle an observer of a quantum system with that system, and so on. These are important and useful concepts when used correctly, but one needs only the basic quantum theory of events, states and probability to understand entanglement.

For those accustomed to thinking in terms of the Schrödinger model, it might be difficult to analyze phenomena using a sequence of quantum events instead of a sequence of quantum states. Of course, I am claiming that all entanglement phenomena can be analyzed as is done here using a sequence of quantum events and their conditional probability. This is a sweeping claim, although it depends on the new definition I have given of entanglement. More specifically, I am claiming that situations that are traditionally called entanglement in the literature are described by Definition 2.8.3. Clearly, this claim will and should be challenged. Let me note that it should be kept in mind when evaluating this claim that pure quantum states are themselves identified with rank 1 quantum events. But do note that in the example of this chapter, there are also quantum events of rank at least 2.

Now, if you insist on using the word 'collapse', that could be acceptable as long as you do not assign a physical significance to it, as long as you do not treat it as a physical process. However, in practice, that does not happen. Many physicists go down a rabbit hole by saying that collapse is a profound topic which requires some deeper 'explanation' or 'interpretation'. Then, it becomes a source of thousands of pages of no significance at all.

Many are so used to working only with the Schrödinger model that they are unaware that some aspects of that model, such as collapse, do not represent physical events. And it is only the probabilistic aspects of quantum events that have a model independent physical significance. Thus, I advocate for discarding 'collapse' from the quantum vocabulary much as the fluid 'caloric' has been discarded from thermodynamical vocabulary. It is not needed. It is best to realize that there is no there there.

Analogies are difficult to find since other branches of science do not rely on two distinct ways of speaking about time evolution. However, biologists sometimes speak of the evolution of species in teleological terms, which they take to be incorrect, yet useful, shortcuts for describing Darwinian evolution. They might say that a certain species of animals evolved to have a thicker subcutaneous layer of fat and white fur *in order* to survive in a colder, snowy climate. This can be easily misinterpreted by those who have not studied evolutionary biology about how environment is related to speciation. But biologists realize that this is just a manner of speaking about how natural selection works. My point is that the two ways of speaking about time evolution in quantum theory (Schrödinger's equation and collapse of the state) are at best a shortcut, a manner of speech, that takes the place of the generalized Born's rule.

3.4 Entanglement without Tensor Products

> Alles sollte so einfach wie
> möglich gemacht werden,
> aber nicht einfacher.
>
> attributed to Albert Einstein

The theory of entanglement presented here does not require representing the underlying Hilbert space as a tensor product. An ample class of such Hilbert spaces is given by $\mathbb{C}^p$, where p is a prime. Recall that the one-dimensional space $\mathbb{C}$ is inadequate for formulating a non-trivial quantum theory since it has only one state and just the two trivial events 0 and I. So, the decompositions $\mathbb{C}^p \cong \mathbb{C} \otimes \mathbb{C}^p \cong \mathbb{C}^p \otimes \mathbb{C}$ do not concern us. In anthropomorphic terms, either Alice or Bob would be trivial. Or in other words, $\mathbb{C}^p$ cannot accommodate both Alice and Bob as non-trivial observers.

The simplest such case is $\mathbb{C}^2$. Of course, this is the well-known setting for studying spin $1/2$ and for defining qubits. But now, this space will be considered in terms of entanglement of events in the von Neumann algebra $\mathcal{L}(\mathbb{C}^2)$. First, here is the usual notation for the three standard *Pauli matrices*:

$$\sigma_1 := \begin{pmatrix} 0 & 1, \\ 1 & 0 \end{pmatrix}, \quad \sigma_2 := \begin{pmatrix} 0 & -i \\ i & 0 \end{pmatrix}, \quad \sigma_3 := \begin{pmatrix} 1 & 0 \\ 0 & -1 \end{pmatrix}.$$

We focus on these particular rank 1 events, which project onto the $+1$ eigenspaces for the three standard Pauli matrices:

$$E_1 := \frac{1}{2}(I + \sigma_1) = \frac{1}{2}\begin{pmatrix} 1 & 1 \\ 1 & 1 \end{pmatrix}, \qquad E_2 := \frac{1}{2}(I + \sigma_2) = \frac{1}{2}\begin{pmatrix} 1 & -i \\ i & 1 \end{pmatrix},$$

$$E_3 := \frac{1}{2}(I + \sigma_3) = \begin{pmatrix} 1 & 0 \\ 0 & 0 \end{pmatrix}.$$

Note that no pair of these events commutes. We continue calculating with the pure state represented by the unit vector

$$\psi = \begin{pmatrix} w \\ z \end{pmatrix} \in \mathbb{C}^2 \qquad \text{where } |w|^2 + |z|^2 = 1.$$

Then, we have these results:

$$P_\psi(E_1) = \langle \psi, E_1\psi \rangle$$

$$= \left\langle \begin{pmatrix} w \\ z \end{pmatrix}, \frac{1}{2}\begin{pmatrix} w+z \\ w+z \end{pmatrix} \right\rangle = \frac{1}{2}(|w|^2 + w^*z + wz^* + |z|^2) = \frac{1}{2}|w+z|^2$$

and $P_\psi(E_3) = \langle \psi, E_3\psi \rangle = \left\langle \begin{pmatrix} w \\ z \end{pmatrix}, \begin{pmatrix} w \\ 0 \end{pmatrix} \right\rangle = |w|^2.$

Next, we have that

$$E_1 E_3 = \frac{1}{2}\begin{pmatrix} 1 & 1 \\ 1 & 1 \end{pmatrix}\begin{pmatrix} 1 & 0 \\ 0 & 0 \end{pmatrix} = \frac{1}{2}\begin{pmatrix} 1 & 0 \\ 1 & 0 \end{pmatrix}$$

as well as

$$E_3 E_1 = \frac{1}{2}\begin{pmatrix} 1 & 0 \\ 0 & 0 \end{pmatrix}\begin{pmatrix} 1 & 1 \\ 1 & 1 \end{pmatrix} = \frac{1}{2}\begin{pmatrix} 1 & 1 \\ 0 & 0 \end{pmatrix}.$$

So, we get these consecutive probabilities for these two events:

$$P_\psi(E_1, E_3) = \|E_3 E_1 \psi\|^2 = \frac{1}{4}\left\|\begin{pmatrix} w+z \\ 0 \end{pmatrix}\right\|^2 = \frac{1}{4}|w+z|^2, \qquad (3.4.1)$$

$$P_\psi(E_3, E_1) = \|E_1 E_3 \psi\|^2 = \frac{1}{4}\left\|\begin{pmatrix} w \\ w \end{pmatrix}\right\|^2 = \frac{1}{2}|w|^2. \qquad (3.4.2)$$

To check for independence of these events (in each order, given ψ), we have to compare these results with

$$P_\psi(E_1)P_\psi(E_3) = \frac{1}{2}|w+z|^2|w|^2. \qquad (3.4.3)$$

In general, the value in (3.4.3) is not equal to either (3.4.1) or (3.4.2). So, with respect to a general ψ, we have that E_1 and E_3 are entangled in both orders. However, for $w = 0$, we have

$$P_\psi(E_3, E_1) = 0 = P_\psi(E_1)P_\psi(E_3) \quad \text{but } P_\psi(E_1, E_3) = 1/4$$

so that first E_3 then E_1 are independent in that order but are entangled in the opposite order. Since ψ is an eigenstate for z spin down in this case, we have $E_3\psi = 0$ and $E_1\psi \neq 0$. So, these probabilities should be intuitively clear for those familiar with spin $1/2$.

Let us note that for two commuting events and a given state, there are two possibilities: either they are independent or entangled. We have seen that both of these possibilities are present in the example of Section 3.1.

However, for two non-commuting events, there are three possibilities:

- They are independent in both orders.
- They are entangled in both orders.
- They are independent in one order and entangled in the opposite order.

In the example above, we see that the second and third possibilities are present in this theory. Next, we give an example of the first possibility. For this, we use the pure state associated to the unit vector

$$\psi = \begin{pmatrix} 0 \\ 1 \end{pmatrix}.$$

One easily computes

$$E_1\psi = \frac{1}{2}\begin{pmatrix} 1 \\ 1 \end{pmatrix}, \qquad E_2\psi = \frac{1}{2}\begin{pmatrix} -i \\ 1 \end{pmatrix}.$$

It immediately follows that $P_\psi(E_1) = P_\psi(E_2) = 1/2$. Next, the consecutive probabilities for the two non-commuting events E_1 and E_2 are

$$P_\psi(E_1, E_2) = ||E_2 E_1 \psi||^2 = ||\frac{1}{4}\begin{pmatrix} 1-i & 1-i \\ 1+i & 1+i \end{pmatrix}\psi||^2 = ||\frac{1}{4}\begin{pmatrix} 1-i \\ 1+i \end{pmatrix}||^2 = \frac{1}{4},$$

$$P_\psi(E_2, E_1) = ||E_1 E_2 \psi||^2 = ||\frac{1}{4}\begin{pmatrix} 1+i & 1-i \\ 1+i & 1-i \end{pmatrix}\psi||^2 = ||\frac{1}{4}\begin{pmatrix} 1-i \\ 1-i \end{pmatrix}||^2 = \frac{1}{4}.$$

So, E_1 and E_2 are independent with respect to ψ in both orders, even though E_1 and E_2 do not commute.

The examples of this section are a way to think about entanglement, even though the Hilbert space $\mathbb{C}^2$ is not a non-trivial tensor product. Despite this being a simple example, it tells us that all three entanglement/ independence possibilities for non-commuting events occur in this theory. This material is familiar from spin $1/2$ physics, but the spaces $\mathbb{C}^p$ for primes $p > 2$ could prove to be more interesting.

3.5 Commuting vs. Non-commuting Events

Let E, F be events. If these are non-commuting events, then in general, their consecutive time order must be specified since the probabilities of EF and FE (in the context of other events) will be different. But if they do commute, then consecutive factors of the two events will give the same probabilities in either time order. In particular, these events can be space-like. This was seen in this chapter where the two commuting events E_1 and E_2 had no defined time order. The moral is that commutating events are not necessarily independent, though they can occur in either possible *consecutive* time order without changing the probability. This may be seen as counter intuitive. However, the probabilities of $E_1 F E_2$ and $E_2 F E_1$ can be different, where F is an intervening sequence of events.

Chapter 4

Schrödinger's Cat

Nothing divided people more deeply
than how they felt about cats.

— Kingsley Amis

But there is another problem with entanglement. It has to do with how
it is verified. Two entangled events E_1 and E_2 can occur with a space-
like separation. In fact, this has been done precisely to see if quantum
theory holds in that case. But how is the comparison of results from a
series of measurements made? After all, at the time of the measurement
neither experimenter can possibly know what the other has measured. So,
they record their measurements and get together at a later time (in the
intersection of their respective future light cones) and compare data. Of
course, such persistent data are not exactly events. This will lead us to
consider the 'paradox' of Schrödinger's cat.

Think of special relativity theory where events in space-time are
described as points in a Minkowski space. Or perhaps, since points are
an idealization, events are small regions in space-time, all of whose points
are very close to each other. One of the conceptual difficulties with relativity
theory is that one discusses objects as if they were events, but somehow
persisting in time. These are called *world-lines*. For example, the physics
professor holds up his piece of chalk for all the class to see. He asks how
long it is. Everyone agrees that it is about 2 centimeters long. But no! That
is only is one inertial frame of reference. The endpoints of the chalk are
actually world-lines of events in Minkowski space and the spatial distance
between these endpoints at the same time (relative to another frame of
reference) is the length of the chalk (in that other frame of reference).
The length in another appropriate inertial frame is 10^{-9} m, a nanometer.
Students are not so easily convinced and rightly so. After all, the concept

of event has been changed behind their backs. And what color is the chalk? Well, the white light of the class room reflects mostly the blue end of the spectrum. All agree that it is a piece of blue chalk. But in another inertial frame the frequency is shifted, and the reflected light from the chalk is red! Even classical physics can be very non-intuitive.

Returning to quantum theory, what is persistent data? I doubt that it is an event which lingers for an indefinitely long time. This seems to be changing concepts, much as in the case of relativity theory. Actually, I have not given any localization information in speaking about quantum events; they are just (time) ordered projections in a Hilbert space. Let me remedy that by using ideas of Haag from his version of local quantum field theory in [18]. As usual everything is in the context of one given Hilbert space $\mathcal{H}$. One thinks of open regions in Minkowski space. For any such region Ω one has an associated von Neumann algebra $\mathcal{V}(\Omega)$ in $\mathcal{L}(\mathcal{H})$. It is well known that a von Neumann algebra has 'lots' (in a sense that can be made precise[1]) of projections, that is, of quantum events. One says that any projection in $\mathcal{V}(\Omega)$ is an event *localized* to the set Ω. Intuitively, one is thinking that it is something that can happen within that region of space-time. These quantum events are the only ones which correspond to something happening in Ω; all other quantum events in $\mathcal{L}(\mathcal{H})$ are not part of the quantum theory in Ω.

Suppose that Ω_1 and Ω_2 are space-like separated regions, meaning that for all $\omega_1 \in \Omega_1$ and all $\omega_2 \in \Omega_2$ we have that ω_1 and ω_2 are space-like events. Under these hypotheses we assume, following Haag, that for all quantum events $E_1 \in \mathcal{V}(\Omega_1)$ and $E_2 \in \mathcal{V}(\Omega_2)$, we have that E_1 and E_2 commute. Of course, the 'small' open sets are the ones that matter, especially those that are small in the temporal direction. To incorporate an event associated to a tubular neighborhood around a very long, especially infinitely long, world-line, into quantum theory is not intuitive for me, since it would be a 'time-extended' event. But I imagine that events associated with these sorts of regions are what one would use in quantum theory to describe the data recorded from scientific experiments or to describe any sort of information that is held indefinitely and can be shared. Of course, the entire scientific enterprise is based on sharing data. And this is what happens also with every entanglement experiment. But in that context it is not considered to

---

[1]For example, the smallest von Neumann algebra in $\mathcal{L}(\mathcal{H})$ containing the events of a given von Neumann algebra $\mathcal{V} \subset \mathcal{L}(\mathcal{H})$ is $\mathcal{V}$ itself.

be a problem. It seems to be 'natural' that the experimenters share their data at some future time. And everyone is in agreement that entanglement has been experimentally verified this way. This is the way one verifies all quantum experiments, actually all experiments in all the sciences! And this is the basis of the often heard statement that quantum theory has been remarkably successful.

But wait! Enter Schrödinger's cat, and the persistence of data from a quantum event becomes a problem. Let's take the innocent cat out of harm's way by changing the set-up. As in the original version, there is a radioactive source which has a probability of $1/2$ of decaying within a certain period of time. A diabolical machine is in place that can detect this decay always and is screened from any possible background noise. (We will see later why it is just as diabolical as Schrödinger's machine.) If the decay is detected, then the machine prints YES on a sheet of paper that is blank except for the experiment run number. Otherwise, it does nothing during the time of the experiment. When the time period ends, the machine ejects the sheet of paper. To check that everything is working well, the experimenter repeats this many times, counts the number of times that YES appears and then divides by the total number of experimental runs. Within statistical precision the relative frequency of YES agrees with $1/2$. But are these sheets of paper quantum events? Is there a self-adjoint operator P whose spectrum is $\{0, 1\}$ with the quantum event $P = 1$ corresponding to YES appearing and $P = 0$ otherwise? If so, then P is a quantum event. And even if this is true, how do the results 1 and 0 persist in time? After all, the data from these events has to be published, lest the experimenter perish. And the sheets of paper can be preserved for future reference an indeterminable number of times for a more or less indefinite future.

Here is my best attempt at dealing with these questions. I think that Schrödinger implicitly assumed that this (quantum!) cat is described by $\mathcal{L}(\mathbb{C}^2)$. Let $P \in \mathcal{L}(\mathbb{C}^2)$ denote the projection which in the cat context tells us whether the cat is dead or alive. This operator generates a commutative von Neumann algebra of dimension 2, namely $\mathcal{V} = \{\alpha I + \beta P \mid \alpha, \beta \in \mathbb{C}\}$ in $\mathcal{L}(\mathbb{C}^2)$. Following Haag the only quantum events available in this quantum theory are 0, P, $I - P$, and I. Note that these are 4 distinct events, since $P \neq 0$ and $P \neq I$ by the construction of the experiment. And only two of these events are non-trivial. Let me note that the event 0 corresponds to starting the experiment with a dead cat, while the event I corresponds to using an immortal cat.

So, what happens now if we take normalized eigenvectors ψ_1, ψ_2 of P, the first for the eigenvalue 1 ('alive') and the second for the eigenvalue 0 ('dead')? We can form the normalized eigenvector (representing a pure state) $\phi := 2^{-1/2}(\psi_1 + \psi_2)$ in the Hilbert space $\mathcal{H}$. But the corresponding quantum event $|\phi\rangle\langle\phi|$ is not in $\mathcal{V}$. In other words, this 'superposition' state of a live cat and a dead cat is not in this quantum theory. It is important to note that this argument does not use a *superselection rule*. Rather it is a mistake to take $\mathcal{L}(\mathbb{C}^2)$ as the von Neumann algebra for this situation. The error lies in thinking that the only von Neumann algebras suitable for doing quantum theory are isomorphic to a Type I von Neumann algebra $\mathcal{L}(\mathcal{H})$ for some Hilbert space $\mathcal{H}$. (This is known as a *Type I error*.) The von Neumann algebra $\mathcal{V}$ is not of this type, since dim $\mathcal{V} = 2$. Rather we have $\mathcal{V} \cong L^\infty(\{0,1\}, \mu_C)$ where μ_C is counting measure, for example. Worse yet, $\mathcal{V}$ is commutative, thereby violating the non-commutativity as required in Axiom 1. And any added event would have to correspond to some physical aspect of the 'quantum cat'. If we only observe the results 'live cat' or 'dead cat', then $\mathcal{V}$ suffices to save the phenomena. However, it is not a quantum theory. We need a theory that includes events for what is observed and nothing more. And even more importantly, we do not–actually can not– give meaning to quantum events that do not correspond to physical results, since it is just such a correspondence which *is* the meaning of a quantum event.

The standard analysis of this 'quantum cat' fails even more spectacularly. Using the states ψ_1, ψ_2 defined above, the state of the 'half-alive, half-dead' cat is taken to be this superposition:

$$\phi := 2^{-1/2}(\psi_1 + \psi_2).$$

(We have been using the usual implicit assumption that the unit vectors ψ_1, ψ_2 are orthogonal.) However, any state in the family

$$\phi_\theta := 2^{-1/2}(\psi_1 + e^{i\theta}\psi_2) \quad \text{for } \theta \in [0, 2\pi)$$

represents a 'half-alive, half-dead' cat, since $|\langle\phi_\theta, \psi_1\rangle|^2 = |\langle\phi_\theta, \psi_2\rangle|^2 = 1/2$. The point here is that two distinct states do not have a unique superposition, but rather an infinite family of superpositions. This fact about superpositions was clearly stated long ago by Dirac in [14]. Yet it is often ignored.

So, which of the states ϕ_θ is the cat 'in' before it is observed? If the state ϕ_θ has any physical significance for a cat (as it does when discussing the spin of an electron), then there should be a physical observable of a cat corresponding to a self-adjoint operator which has ϕ_θ as a non-degenerate eigenstate. It is not clear at all what any of these operators has to do with a cat, though they make sense as observables for the spin of an electron. The corresponding events $|\phi_\theta\rangle\langle\phi_\theta|$ have no relevance for the cat and have been excluded above from the associated von Neumann algebra.

It gets worse. Probabilities that make perfect sense for the spin of an electron give non-sensical results for the cat. It does not even make sense to assign states to 'alive' and 'dead' as done above and as is typically done when discussing this topic. The usual assumption $\langle\psi_1,\psi_2\rangle = 0$ implies that a live cat, taken to be an isolated system, has zero probability of dying. Of course, a cat completely isolated from its environment, including air, will quickly perish. And if $\langle\psi_1,\psi_2\rangle \neq 0$, then an isolated, dead cat has non-zero probability of becoming alive. The von Neumann algebra $\mathcal{L}(\mathbb{C}^2)$ works well for the spin of an electron, but fails miserably for a cat. Assuming that the knowledge of Wigner's friend is also in one of the states ϕ_θ (but which one?) only adds even more confusion. There is no known quantum theory of such complicated biological entities such as cats, Wigner's friends and human knowledge, given current scientific understanding. For a cat or indeed for any biological organism this would include quantum constructs corresponding to 'alive' and 'dead'. Unsurprisingly, the toy quantum model for a cat as presented above does not work. This is not a paradox, but just another indication that quantum theory has yet to describe all physical phenomena.

Moreover, it seems to be folklore that 'one can not write something down on paper in quantum theory'. This is related to the no-clone theorem of quantum theory. (See [33].) But my diabolical machine does write something down on paper. And this is why it is diabolical. It defies quantum theory. What are we to make of that? Some physicists hold to the view that all physical phenomena are described by quantum theory. So, they face the challenge of describing in the context of quantum theory how data can be recorded and shared. It is not permitted to appeal to classical physics in such an argument. It has to be a demonstration totally within quantum theory itself. This is what I think is the real *Measurement Problem of Quantum Theory*. Other physicists assert that the diabolical machine as well as all measuring devices are classical, not quantum, systems. I will examine this problem further at the end of this section and in the next chapter.

Nonetheless, I wish to note that there do seem to be recording devices in nature that are usually considered to be quantum systems. Let us think of a nucleus that is unstable via beta decay. The nucleus changes from one element in the periodic table to another when it decays. This is recorded in many ways; for example the electric charge of the nucleus changes. So, the electric charge serves as a way of recording whether the nucleus has decayed or not. And if the daughter nucleus is stable, then this data is recorded for all posterity. Nuclei are certainly not classical systems. And most physicists would agree that they are quantum systems. So, this is an example of how some data (though possibly not all data) can be recorded in a quantum system. The wheel has come full circle. It was a radioactive decay that initiated the events of the diabolical machine. In short, forget about the diabolical machine, the box, and the cat inside the box. None of that needs any more explaining than how, according to quantum theory, data about beta decays can be *measured* and then *shared*. So, we find the same puzzle at the end of the story as at the beginning, but now in a quantum system.

The same problem is present in classical physics, though it seems to never be discussed. It is never simply a question of a physical system and some observation of it. Rather, the information from that situation is recorded and shared. The question is: 'Can the physical processes of recording and sharing information be explained using classical theory?' The answer is not likely to be yes. Classical mechanics describes motion as a trajectory in a phase space. The enormously high dimensional phase space of a book, say, has nothing to do with the information printed in it using black ink on white paper. After all the differential reflectivity of ink and paper are explained only by the quantum theory of the interaction of light with matter. This particular technology was the basic method for recording and sharing scientific information for millennia and is still used to this day. More recent technologies based on electromagnetic storage devices are also only explicable using quantum theory, not classical theory.

My point is that the physics community is not agitated when this serious question is not addressed in discussions of classical mechanics, for example in a university course or in a popular exposition for a general public. Typically, one hears that there is no measurement problem in classical physics, while there is such a problem in quantum physics. What's sauce for the goose is sauce for the gander.

Chapter 5

Measurement Problems

Truth is truth to the end of reckoning.
Isabella in *Measure for Measure*

— William Shakespeare

A measurement problem has already appeared in the previous chapter. It is not a problem that one can address, let alone solve, in the context of an axiomatic theory. Let me illustrate this with an example which I hope is not very controversial: Euclidean geometry. In the full axiomatization of this theory (which eluded Euclid's efforts) certain statements are proved about the size of angles and the areas of geometric figures, such as squares in the Pythagorean theorem. These statements are in contradiction with statements in non-Euclidean geometries about the 'same things'. It is generally accepted that a way to test if Euclidean geometry holds is to measure the three physical angles of physical triangles and see whether their sum is indeed π radians (in modern units). It does not pertain to Euclidean geometry nor to any of its competitors to explain how to measure angles or areas. That is not understood to be part of the task of the theory itself. This question is not considered to be one which the theory has to answer: How does one measure an angle? In fact, one assumes that physical triangles with three physical angles exist and, moreover, that all of these incompatible geometries are speaking of these same physical objects. There are no Euclidean angles, no projective geometry angles, and so on. But there are measured angles. Similarly, there is no classical energy, no quantum energy and so on. But there is measured energy. For an experimental scientist measurement is an activity, not a problem.

And I see the axiomatization of quantum theory in the same way. As presented in this treatise quantum theory is an axiomatic theory of states, events and probabilities. How these are measured is not addressed in the mathematical theory. (But I do come out in favor of relative frequency as the way to measure probability.) One should think of 'state, event, probability' as three basic concepts much like 'line, angle, area' in geometry. Of course, these latter geometric words get significance in terms of observations. And so do the former three concepts of quantum theory. But those are scientific questions outside the scope of the axiomatic theory. Alternative approaches to quantum theory must have different axioms for 'state, event, probability' which then must lead to different consequences that can then be checked by experiment. (If the consequences are not different, we would simply have a reformulation of the given approach.)

But I think that what is usually meant by the Measurement Problem in Quantum Theory is part of the theoretical approach, having to do with the collapse condition. Besides being an arbitrary and seemingly non-intuitive rule, collapse introduces a second way that states can change in time.

- The first way is via the solution to Schrödinger's equation. This change is continuous (in an natural topology) and deterministic, meaning that given an initial condition there is a unique solution for all times.
- The second way is via collapse, which is induced by measurement and which is both discontinuous and probabilistic. (In the simplest case, the collapse changes the initial state to the equivalence class of the normalized eigenvector of the probabilistic non-degenerate [eigen]value that was measured.)

Moreover, the first way is known once the Hamiltonian H of the system is known, and H is an *intrinsic property* of the system. On the other hand the second way depends on an *external measurement* of the system. While this is a much discussed problem, I think that this is how the usual *Measurement Problem* enters quantum theory, namely measurements that are external to the system govern its time evolution as much as the intrinsic dynamics given by its Hamiltonian. Or as it is sometimes phrased, quantum theory has two different time evolutions.

It is important to pause here and note that I am defining *a* Measurement Problem here. And I think that this is an accurate formulation of the *usual* Measurement Problem as it appears in the literature. While that literature is vast and opinion may differ, in order to solve a problem one must define it first. If others think this is not an adequate formulation of *their* Measurement Problem, then it behooves them to define what that is. Also, note that *this* Measurement Problem is a problem in the usual quantum theory, not in experimental practice.

Worse yet, there is no other physical theory which admits two distinct time evolutions. In the approach of this treatise, the second way above is not physically meaningful, but is only one step in a two step algorithm for computing one quantum conditional probability, which is itself physically meaningful. Also, in this treatise the first way above is not the principal time evolution in all models, although it is behind the time dependent Generalized Born's Rule in the Schrödinger model. So, I claim that this approach solves the theoretical Measurement Problem as defined here.

However, the question remains how to relate experimental measurements to this approach to quantum theory based on states, events and probability. Appeals to classical physics, or any other theory, are beside the point. What is sought for is a consistent, but not circular, exposition of the topic. So, I will not solve this *experimental* Measurement Problem but rather explain what are some issues that a scientific approach should address. And in fact, I am quite unprepared for solving *this experimental* Measurement Problem. I am only trying to put experimental measurement into the context of this treatise in terms of states, events and probability. However, I do think the theoretical Measurement Problem, as defined above, is now resolved by the clarification of the roles of Schrödinger's equation and of collapse in terms of the Generalized Born's Rule.

To start off I would like to introduce the idea that measurement is done by devices (or physical systems, if you wish) that implement sequences of two events in such a way that the conditional probability of the second event, given the first event (and maybe a state) is 1. It is definitely not assumed that such a device always has the same sequence of two events. Such a device is essentially a function that has a certain set of events as the possible first event (input) of the sequence and then has the second event (output) in that sequence. It should further be required that this function is one-to-one, so that the second event in the sequence uniquely specifies the corresponding first event. Colloquially, the second event is characterized by the first event, and it in turn uniquely characterizes the first event.

Let me emphasize this one more time. Sometimes the claim is made that a measuring device must be classical since its results are 'determined'. The assertion is made that this is a deterministic process, and hence the device can not be a quantum system but must be classical system. But this logic is fallacious. Probability theories include the special case of conditional probability 1, as we have seen in Chapter 3 on Entanglement.

Whether such devices actually exist physically is another question, though I will assume that they do. Anyway there seems to be nothing in quantum theory which precludes such devices from existing. Measuring devices as defined here should be consistent with quantum theory. And sequences of two events that have conditional probability 1 are consistent with a probabilistic theory; they are just a special case. How to explain particular physical devices using quantum theory is something that depends on the specific details of those devices. Though I doubt there is a general answer for all devices, I could be proved wrong. In short, I suspect that giving a general quantum theory of measuring devices is an intractable problem, at least with our current level of understanding. This seems to be born out by the overwhelming lack of progress in addressing this issue. But for those who think otherwise and want to research this problem, please remember the important role of approximation. The point is that devices which implement sequences of events with probability nearly equal to 1 can be quite accurate, though not perfect, measuring instruments.

Quantum computing can be viewed in the same way. Though speaking of a sequence of states being changed by quantum operations is valid, that language does not survive in the equivalent Heisenberg model where the state is time independent. I would rather say a hard-wired quantum computer is a physical device that has a final (output) quantum event that is produced *with high conditional probability* by a given first (input) quantum event and any state. Only in this case, the function being implemented this way need not be one-to-one; distinct input could result with high probability in the same output. A programmable quantum computer would do the same, again with high probability, but the function would depend on a program. To my understanding this is what 'classical' computers do, but with a set of events that is a Boolean algebra. A fully quantum computer would use all quantum events.

Let's look at Schrödinger's cat using these ideas. A radioactive nucleus with charge Z is under consideration. Let E_1 be the (quantum) event that has value 1 if and only if beta decay occurs in some time interval. Let E_2 be the (quantum) event that has value 1 if and only if the cat dies in the same time interval. The device is constructed so that for any state ψ we have that the following conditional probabilities hold:

$P_\psi(E_2 \mid E_1) = 1$ given that decay occurs, cat dies,

$P_\psi(E_2^c \mid E_1) = 0$ given that decay occurs, cat does not die,

$P_\psi(E_2 \mid E_1^c) = 0$ given that decay does not occur, cat dies,

$P_\psi(E_2^c \mid E_1^c) = 1$ given that decay does not occur, cat does not die.

Here, $E^c = I - E$ is the complementary event to the event E. For example, the event E_1^c means that the beta decay did not occur in the time interval. It is not to be confused with the expression 'nothing happened' in ordinary language, which misses the point. No one, starting with Schrödinger himself, ever seems to doubt the possibility of constructing a device which has these conditional probabilities.

Seeing a live cat is a way of measuring the charge of the nucleus and getting the result that it is Z. And seeing a dead cat is a way of measuring the charge of the nucleus and getting the result that it is $Z + 1$. Of course, there are other ways of measuring the charge of the nucleus. One might say they are more 'direct' than using the poor cat as the measuring device. But I claim that they will also be based on sequences of quantum events which have conditional probabilities equal only to either 0 or 1. Or maybe just with conditional probabilities only very near 0 or very near 1.

While this may be rather clear, it conceals what I do consider to be the fundamental problem with all measurements: how is measurement converted into reliably, though not necessarily perfectly, reproducible information. In the example of Schrödinger's cat, this has to do in the previous paragraph with the word 'seeing', which unfortunately introduces an observer into the discussion. However, somehow or other human observation — and even more importantly — communication are relevant. The problem is how to explain with physics the existence and properties of information. Classical physics was started and prospered in an age when information was preserved only by writing, both in natural as well as in mathematical language. This is simply not explicable in terms of classical physics. Dark ink and white paper (as all colors of physical objects) are properties of the interaction of light with matter which is explicable only in terms of quantum physics, not classical physics. Then the question is if quantum theory can do better. This goes beyond the problems of how to construct computers that are fundamentally quantum systems. As I see it, the fundamental Measurement Problem is how to use physics to explain the existence and properties of information as we produce, copy and use it in all aspects of social life and not only with respect to scientific measurements. This is much more general than understanding the role of measurements of quantum systems. One controversial proposal is to turn this problem on its head by taking Information Theory as being the fundamental theory and then using it as the basis of physics. For various ideas for doing this see [9], [13] and [43].

Chapter 6

The EPR Paper

Face au réel, ce qu'on croire savoir
offusque ce qu'on devrait savoir.

— Gaston Bachelard

Ever since its publication in 1935 the famous EPR paper [15] (named for its authors Einstein, Podolsky and Rosen), has generated much discussion and controversy, despite the fact that it is basically correct. It has even been referred to as the EPR paradox! The authors come in the last paragraph of [15] to the following conclusion:

> While we have thus shown that the wave function does not provide a complete description of the physical reality, we left open the question whether or not such a description exists. We believe, however, that such a theory is possible.

The argument of this paper is an example of entanglement, though that word is not used, presented in the Schrödinger picture. To achieve their conclusion the authors have to give some idea of their concept of 'physical reality'. This they do by saying the following:

> *If, without in any way disturbing a system, we can predict with certainty (i.e., with probability equal to unity) the value of a physical quantity, then there exists an element of physical reality corresponding to this physical quantity.*

(Italics in the original)

They then add how to regard this condition:

> Regarded not as a necessary, but merely as a sufficient condition of reality...

One can quibble with their condition, but for the moment let's accept it and try to understand the paper a little better. At the beginning of the paper the authors state:

> In attempting to judge the success of a physical theory, we may ask ourselves two questions: (1) 'Is the theory correct?' and (2) 'Is the description given by the theory complete?' ...

A little further on they continue:

> It is the second question that we wish to consider here, as applied to quantum mechanics.

This is also the question posed in the title of the paper. However a careful consideration of both their argument and of their very words shows that they did not answer this second question. Just to be clear here again is the essential part of their correct conclusion:

> ... the wave function does not provide a complete description of the physical reality ...

So, their implicit assumption is that the wave function is the only theoretical element that is involved in the description of quantum theory. That is how they can jump from showing the incompleteness of the wave function to the incompleteness of quantum theory as a whole.

Of course, this is an error of logic known as a *non-sequitur*. But it is easy to see how the authors arrived at it. Apparently, their tacit assumption is that the Schrödinger equation is the fundamental time evolution equation of quantum theory and that equation has a unique solution (the wave function), given an adequate initial condition. This is second nature for those accustomed to thinking in terms of physical theories based on differential equations; this is seen as intuitive. Unfortunately, as we have seen, it is wrong. The details in the course of their argument about how the state changes due to measurement are not viewed as contradicting this general idea of how physical theories work. Of course, these details are nonetheless just how one calculates conditional probability in the Schrödinger picture, though that wording is not used.

However, we can suppress this tacit assumption and see that the actual conclusion of the paper is correct. Indeed, the wave function alone does not suffice to provide a complete description. As presented in this treatise, events and probabilities (and their possible time evolution) are also part of basic quantum theory. It might well be that the concept and importance of quantum event were not known to the authors of the EPR paper in 1935. But probability was a contemporaneously available concept. Indeed, probability one is singled out for special consideration in their sufficient condition for an element of physical reality. However, later commentators, and perhaps these authors on later occasions as well, have turned this into a necessary condition if they assert that something with probability less that one is not an element of physical reality. While one may wish to take their condition as both necessary as well as sufficient, that is a giant step beyond what the authors of EPR actually assert in their paper.

Some confusion arises since rank 1 events are pure states ('wave functions' for these authors). But we saw that entanglement in the simplest example of Chapter 3 involves events that are rank 2 operators and therefore not wave functions. Similarly, the EPR thought experiment involves events which are not wave functions. This may seem to be just a minor technical detail, but ignoring it leads to a semantic flaw in the EPR argument. Using the notation of Chapter 3, the occurrence of E_1, a rank 2 event, does not tell us the state of the particle on the left in the Hilbert space $\mathcal{H} = \mathbb{C}^2 \otimes \mathbb{C}^2$, a 4-dimensional space. As shown in Chapter 3 entanglement can easily be understood by calculating the conditional probability of events without assigning 'meaning' to the intermediate steps. But if one prefers the language of states, then the correct statement is that the occurrence of the event E_1 tells us that the state is in the 2-dimensional subspace Ran E_1. Similarly, E_2 does not tell us the state of the particle on the right. The implicit error here is the assumption that the measurements of the detectors are being modeled by events in two different 2-dimensional Hilbert spaces. So, one is discussing this quantum system with three distinct Hilbert spaces, and therefore three different von Neumann algebras. This is a clear violation of Axiom 1. In short the EPR thought experiment does not assign, by using the authors' sufficient condition, an 'element of reality', namely the state of either particle.

Let's examine this in a bit more detail. To describe the detector on the left we only need the von Neumann algebra $\mathcal{L}(\mathbb{C}^2)$. A different copy of the same von Neumann algebra suffices for the other detector. But if the space-like separated events of the detectors are compared and found to have some correlations among them, then a scientific puzzle has been noted that requires further explanation. Of course, we already know what that explanation is in this case. And a part of that is excluding the two copies of the von Neumann algebra of the individual detectors and considering instead $\mathcal{L}(\mathbb{C}^2 \otimes \mathbb{C}^2)$, the new von Neumann algebra for the combined system.

Let's take this one step further and suppose that the detector on the left not only says that E_1 occurred, but the experimenter on the left is told that the earlier event E_0 had already occurred. To accommodate the event E_0 would require using a new model, the simplest one of which would be the von Neumann algebra $\mathcal{L}(\mathbb{C}^2 \otimes \mathbb{C}^2)$. Using this model the experimenter on the left then has sufficient information for calculating the *probability* of the event E_4 of any spin $1/2$ measurement by the detector on the right. As I will argue in a moment, this is sufficient for asserting that the value ($\equiv$ eigenvalue) of the event E_4 has an element of physical reality.

Now, I wish to suggest that their sufficient condition for element of physical reality is unnecessarily linked to probability one events. Leaving aside the vague restriction about not disturbing the system, I propose another more general, sufficient condition for an element of physical reality:

If we can predict with non-zero probability the value of
a physical quantity, then there exists an element of
physical reality corresponding to this physical quantity.

I might even be disposed to deleting the word 'non-zero' from the above condition. After all, violation of a conservation law has a physical reality of a certain sort — the reality of never occurring (to date). This condition seems to be more in the spirit of basic quantum theory viewed as being a new type of probability theory. Of course, in some sense I am proposing a broader definition of 'element of physical reality'. However, this condition is intuitive in the sense that people act as if it holds. For example, a specific house burning down and resulting in (financial) value zero has an element of physical reality due to its non-zero theoretical probability, even though it has never happened and most likely never will. Yet this element of physical reality is sufficient to motivate buying fire insurance for the house.

Anyway, with this enhanced sufficient condition physical quantities such as position or spin component are elements of physical reality. Moreover, such physical quantities do find a counterpart in quantum theory; their pvm's and probabilities are the counterparts. This would make quantum theory nearer to being complete than the authors might have thought according to their own criterion given in the EPR paper:

Whatever the meaning assigned to the term *complete*, the following requirement for a complete theory seems to be a necessary one: *every element of the physical reality must have a counterpart in physical theory*.
(Italics in original)

The authors of the EPR paper include as an element of physical reality any probability 1 physical event. Its counterpart in the quantum physical theory is the theoretical probability as given by the Generalized Born's Rule, which better give the value 1 as well. The wave function (or state) alone is not sufficient for calculating that theoretical probability by using the Generalized Born's Rule. All of the pertinent events must be included as well.

I am rather befuddled by those who speak of the EPR paper as a paradox, especially since that word is never used in that paper. And after all, there is nothing paradoxical in asserting that a theory is incomplete. The only thing paradoxical about the EPR paper is that some think it is paradoxical. However, it is a poorly written paper which arrives at a correct assertion and then proceeds to an incorrect conclusion. Clearly, quantum theory remains incomplete but for other reasons such as its current inability to predict the values of fundamental constants, the nature of gravitation and the origin of the universe to name some examples, none of which is a paradox.

Chapter 7

Determinism and Probability

> The epistemological value of probability theory is based on
> the fact that chance phenomena, considered collectively
> and on a grand scale, create non-random regularity.
>
> — Andrei Nikolayevich Kolmogorov

There is a probabilistic aspect in all the standard formulations of quantum theory. There is rarely a probabilistic aspect in standard formulations of classical physics and, when there is, it is chalked up to being due to a lack of complete information rather than a fundamental aspect of that theory. But it is also said that classical physics is deterministic. And this is not a logical consequence of what I just said. Rather it is known as jumping to a conclusion. The determinism ascribed to classical physics is boilerplate repeated over and over by so many experts that it has become heresy to question it. Note that a mathematical formulation of determinism is not found in the standard texts on classical physics. The best one can find in the scientific literature are various theorems in mathematics which state that some classes of time dependent differential equations have unique solutions provided they are accompanied by appropriate initial conditions. But it turns out that there are other time dependent differential equations that do not have this property and, according to theorem, such equations are non-linear. Also there is nothing in classical Newtonian mechanics which tells us which sort of differential equations will arise in that theory, that is, whether they have unique solutions with appropriate initial conditions or not. But it is known that the second law of motion of Newtonian mechanics is non-linear in almost all examples.

Since this is the relevant moment, let me pause from continuing this discussion in order to call out a confusion that is rampant in the physics community. The point is that differential equations do not automatically imply determinism. To suppose otherwise is wrong-headed, to say the least. Some differential equations are consistent with determinism, but others are not. We often work with a linear approximation. Linear differential equations with appropriate initial conditions do have unique solutions, which are also global in time. But this property of a particular approximation does not always hold in other non-linear contexts. So, it is a *non-sequitur* to say that a theory based on differential equations is automatically deterministic.

Let me be clear about this difference between classical mechanics and quantum theory. The theory of classical mechanics does not answer the question of whether determinism or probability applies to its topics. On the other hand quantum theory at its most basic level makes statements about probability and, according to this treatise, about nothing else testable[1]. One should not say that quantum theory eliminated determinism from physics; rather it put probability into physics. And probability is very non-intuitive.

However, I pause again, since it appears that Schrödinger's equation is deterministic. After all, for any self-adjoint Hamiltonian H there is a unique solution, global in time t, given any initial state φ in the Hilbert space, namely, $\psi(t) = e^{-itH} \varphi$ in the Schrödinger model. At a purely mathematical level, this is a sort of determinism. But this does not give us determinism of a quantum system, since the states $\psi(t)$ by themselves only determine probabilities when combined with the events associated to a system. And this role of probability is not the lockstep relation of cause and effect as envisioned by the usual idea of determinism. The Heisenberg model also has a mathematical determinism, but now of the events as given by Heisenberg's equation (1.7.3). And again this is not a physical determinism, since only probabilities (and not definite physical outcomes) are determined.

And moving on now to experiment the situation is even shakier. The idea is that by knowing all about a physical system completely at one instant of time, the future of that system is known or, as one says, determined. How could one possibly know that? How could one possibly falsify that? If some systems behave in what seems to be a deterministic manner (at least for some experimentally finite period of time), that does not mean they will continue to behave in that manner forever. And if they do not behave in a deterministic manner, it is unacceptable to say that is due to lack of knowledge about the initial situation. How could one possibly know *in all cases* that there is more to know? It is logically possible that there is nothing

---

[1] Meaning verifiable or falsifiable, according to your preference.

more to know. There is a lot of muddled thinking about determinism. Of course, there is a lot of muddled thinking about probability theory, too.

Something that is truly confusing for me is that some scientists cannot think about probability 1 quantum situations without invoking deterministic language. For example, entanglement is often described deterministically. However, such situations can and should be explained in terms of quantum probability, not with determinism. Following the quotation of Kolmogorov, probability 1 is just one example of non-random regularity. One really must carefully distinguish a probability theory in which some of the probabilities are 1 while others lie between 0 and 1 from a deterministic theory in which one can assign probabilities as long as they are either 0 or 1. Viewed this way one sees that determinism is just a special, degenerate case of probability.

From this perspective it is not so strange to say that quantum theory is probabilistic and, in fact, this is thought to be one of the basic aspects of the theory. I have done just that in this treatise. What is strange is that a deterministic time dependent differential equation is claimed to be another basic aspect of quantum theory. Of course, I refer to the time dependent Schrödinger's equation

$$i\frac{d\psi}{dt} = H\psi. \tag{7.0.1}$$

Here in more detail is what the mathematical theory actually says about this equation. We assume that H is a self-adjoint densely defined linear operator acting in a Hilbert space $\mathcal{H}$. Then a theorem says that for any ϕ in the dense domain of H there exists a unique solution ψ_t of (7.0.1) for all $t \in \mathbb{R}$ such that $\psi_0 = \phi$. Moreover, that theorem asserts that $\psi_t = e^{-itH}\phi$, where the globally defined unitary operators e^{-itH} are defined for all $t \in \mathbb{R}$ by the functional calculus of spectral theory applied to the operator H.

In the Schrödinger model one has Schrödinger's equation, a deterministic equation, and a probabilistic state collapse condition. It seems that the time evolution chugs along on its merry way changing the state is a continuous deterministic way, when–poof!–somehow a probabilistic event occurs that changes the state in a discontinuous way. How can there be two distinct types of time evolution? This perennial puzzle about quantum theory arises as we have seen by focusing on details in the Schrödinger model which are model dependent. Of course, in the equivalent Heisenberg model there is a deterministic time evolution of the pvm's while the state remains constant in time. In both models the (identical!) probability measures are continuous functions of time. The continuity of these probability measures follows from the continuity of $t \mapsto e^{-itH}$ in the strong operator topology.

For now I would like to discuss another sort of probability in one aspect of quantum theory. But this aspect is quite different. Again, consider a nucleus that can undergo beta decay. Before the decay occurs the electron and anti-neutrino of the final state do not exist. They are created by the decay process. And no matter where the nucleus is in the universe, the electron has exactly the same characteristic properties as any other electron in the universe. Its mass, electric charge, spin and magnetic moment are always the same. In other words these are probability one properties of any electron. The same holds for the anti-neutrino; it is the same as any other anti-neutrino produced in beta decay. To make this sound more 'paradoxical' imagine two beta decays of the same isotope nucleus, but with a space-like separation between these two events. How can one event possibly 'know' about the other? Yet they produce the same decay products, although the momenta of the decay products are not always the same. Also, these decays violate parity conservation, but they do that in the same way always. On the other hand an alpha decay of a nucleus always preserves parity. These are all rightly called *particle properties*, though one could also describe them as *spooky action at a distance*.

Chapter 8

Philosophical Questions

Se moquer de la philosophie
c'est vraiment philosophe.

— Blaise Pascal

Any human activity — economic, judicial, scientific — is a valid topic for philosophical discussions as to what are the intellectual principles that are implicitly being used. Nonetheless, the philosophical questions raised by quantum theory arrive at considerations not usually seen in other contexts, not even in other scientific contexts such as biology. Though far from being an expert on these matters, my perspective from afar is that the central role of probabilistic analysis in quantum theory is the source of the more perplexing questions.

The history of philosophy has had much weight it seems to me. I fear that this has overshadowed the analysis of some philosophical questions relevant to quantum theory. Certainly, the precursors and the successors of Aristotle play an important historical role. To a great extent that tradition produced the very language we use to this day in order to understand what a 'scientific explanation' is. A basic idea in that language is given in terms of *cause and effect*. The simplest expression of this idea is that there exist causes in the natural world, and these lead inevitably to their uniquely specific effects. One says that a cause *determines* its effect. In this sense the cause is the explanation of its effect and perhaps in some contexts the word 'explanation' might be a better choice than the word 'cause'.

In other words scientific activity can be simplified by saying observations are made of effects and then experimentally verifiable causes of these are sought. For example, if one hears thunder, then one seeks its cause, which turns out not to be so simple to do. Once as a child when thunder frightened me, I was reassured that it was just the sound of clouds bumping into each other. It appears to be an aspect of human nature to seek explanations (or causes). Knowing whether any given explanation is 'true' is its own complicated and separate issue, as the example of thunder shows.

How does quantum theory as presented here fit in with this cause and effect way of explaining phenomena? Well, the simplest answer is that it does not, but rather offers a new way of explaining that at a fundamental level is not that of a well defined cause determining its unique effect. That is to say, in this axiomatic approach to quantum theory, the relations between events are explained by probabilities, nothing else. This is difficult, if not impossible, to understand for those accustomed to cause and effect analyses. For example, it leads to misleading statements such as:

An electron can be in two places at the same time.

As already seen in Section 2.7, this is a confusing way for saying:

An electron can have non-zero probability to be
at each of two places at the same time.

At the level of theoretical explanation as given here, only such assertions about probabilities can be made. It remains as a problem to scientifically decide the truth or falsity of these two distinct statements. Of course, if the first assertion can be empirically supported, then this approach to quantum theory will be falsified. And, as an aside, conservation of electric charge will also be falsified.

As is well known a probabilistic formulation motivates[1] further questions about what is 'real'. However, within this theoretical formalism there is no assertion that anything whatsoever — not even events — are real. While this seems to fly in the face of the scientific project of explaining the real world, it is not as bizarre as it might first seem to be. Let me return again to another purely mathematical and axiomatic theory to illustrate this thesis: Euclidean geometry. That is a theory of points, lines, circles and so on. It may or may not be adequate for describing physical space, but note that these theoretical concepts do not exist in the physical world. Or more precisely, they do not need to exist in order to apply Euclidean geometry in practical applications such as land surveying. At the very least it must be admitted that points have never been observed. The same goes for lines,

---

[1] Note the human connection!

circles and so on for all 1-dimensional figures. Yet we can say — and can
agree — that a beehive has a hexagonal structure. This goes to the heart
of historically important scientific revolutions, such as that concerning the
nature of the movement of the planets. Do they move in circles (with added
on epicycles) or in ellipses? But 3-dimensional objects, such as planets, do
not move along 1-dimensional curves of any type. These comments are not
intended to diminish the importance of Euclidean geometry. They merely
are meant to show that a useful theory can be based on concepts that do
not have a physical reality.

This treatise takes the basics of quantum theory to be certain probabilistic statements. The title question of the EPR paper can then be changed to:

Is the probabilistic description of quantum theory complete?

As such, this question can clearly be answered scientifically in the negative
by providing an alternative formulation which describes more, that is, which
is experimentally better. But to answer this question scientifically in the
positive seems to me to be beyond reach. However, here is an independent
question in the philosophy of science in general:

Can a probabilistic description of any scientific theory be complete?
And, if so, what are sufficient criteria for doing so?

I think that in many examples probability theory is what gives quantum
theory its non-intuitive character. But probability seems to play a role as
well in some philosophical questions concerning quantum theory. This could
merit further reflection.

Another philosophical question: What does random mean for physical
phenomena? That is, what is the meaning of *randomness* in the physical
world? For example, it has been maintained that probability, a mathematical concept, does not exist in the real world. That's all well and good,
since it is in resonance with Euclidean geometry. But something in reality,
commonly called randomness or chance, must somehow correspond in some
sense with mathematical probability. So, the question is: How does all that
work? While I have proposed relative frequencies as a practical answer to
this question, it could also have philosophical aspects, which I leave to the
experts.

Chapter 9

A Quaternity of Problems

Avant tout il faut savoir poser des problèmes.
Et quoi qu'on dise, dans la vie scientifique,
les problèmes ne se posent pas d'eux-mêmes.

— Gaston Bachelard

Isham poses these 'conceptual issues in quantum theory' as a 'quaternity of fundamental problems' in [21]:

1. The Meaning of Probability.
2. The Role of Measurement.
3. The Reduction of the State Vector.
4. Quantum Entanglement.

Even though all four of these topics (now viewed as problems!) have been previously presented, here is a resumé of all that.

For the first of Isham's problems, the best I can say is that quantum probability is the one invariant feature of the various 'pictures' of quantum theory. In particular, probability does not have any physical importance for single events, since science explains correlations. So, some of the confusion about the meaning of probability has to do with the focus on the wrong theoretical concept. And these quantum probabilities are meant to explain the relative frequency of *multiple* physical events. That is their meaning and is what the *time dependent Generalized Born's Rule* deals with. Also this is the *one and only time evolution equation* of quantum theory, thereby giving it the central role and meaning in quantum theory. The Schrödinger equation is just a special, and important, part of one model. Therefore, this approach is *fully compatible* with standard quantum theory and explains why 'shut up and compute' works. The moral is that Quantum Probability is important in all its aspects: consecutive and conditional probability, not just the probability of a single event. Failure to consider all of these probabilities is a major shortcoming of the introductory quantum texts, including my own text [39]. For some people it still may defy their intuition that probability is a fundamental aspect of quantum theory, but that is not a valid criterion for rejecting a scientific theory. However, the approach of this treatise does not preclude the possibility of finding some more basic explanation of how probability theory arises in quantum theory. If so, that might give added meaning to quantum probability. What I have said here may not provide the meaning of probability to everyone's satisfaction but nonetheless it should clarify how it functions.

I have even ventured to go so far as to say that single event probabilities are irrelevant. And this is because *science is based on the understanding of correlations*, not on individual, unrelated events. Isham's four problems are a case study of what happens when this principle is not heeded.

Measurement does not have a special role in this approach to quantum theory. Rather, in this theory measurements are simply a certain type of quantum event, namely those of the form $P_A(B)$, where A is a self-adjoint operator and B is a small interval in $\mathbb{R}$. But we did not need to explicitly discuss these sorts of quantum events separately. Rather the empirical fact that measurements generate information, which has its own properties, is a more fundamental Measurement Problem that current physics does not explain. What 'properties' are really being measured is not an issue in this approach, though it is an important consideration. Recall that this treatise is not intended to be a complete rendition of quantum theory. However, the usual theoretical Measurement Problem as defined in Chapter 5 has been resolved by the Generalized Born's Rule.

The 'reduction of the state vector' (or 'collapse') is not seen here to be a problem. It is simply one step in a two step algorithm for computing a conditional probability. As such it really does not merit having a special name nor any special consideration. And it is certainly not an event as events are understood in this approach.

Lastly, quantum entanglement has been given here a new, more general, definition as the lack of independence in terms of the theory of quantum probability of multiple events in the context of one von Neumann algebra. This clarifies most previous discussions, which attempt to explain this in terms of only single events in multiple Hilbert spaces. This elucidates its meaning, though again it may not satisfy everyone's intuition. Speaking for myself only, I find quantum probability difficult to grasp intuitively, although the required calculations are rather straightforward.

It is important to emphasize that correlations (i.e., entanglement) only can refer exclusively to what occurs, namely multiple events. There is no entanglement of anything else, such as observers, reference frames, measuring devices, or anthropomorphic characters (say, Alice, Bob, etc.) or various animals, human and non-human. Measurements, being a special type of event, can be entangled. A pure state ψ can be considered entangled in the sense that we have the associated projection operator $|\psi\rangle\langle\psi|$, which (if it is in the von Neumann algebra $\mathcal{V}$ of the quantum system) can be entangled (i.e., correlated) with other events in $\mathcal{V}$.

This clarifies much of the confusion surrounding these four problems. This has been achieved by using this groundbreaking approach, which includes probabilities of multiple events instead of limiting the theory to only the probabilities of single events or single measurements. However, a new and more fundamental Measurement Problem has now been posed: How can physics explain the existence of reproducible, long enduring information.

Chapter 10

Interpretation

Interpretation is the revenge of intellect upon art.

— Susan Sontag

The astute reader will have noticed that I have only used the phrase *physical significance* instead of 'interpretation' as was done in days of yore for describing the relation between theory and observation. This is due to the unfortunate situation that 'interpretation' has come to denote something quite different from 'save the phenomena', a phrase also from days of yore. (See [35].) Nowadays 'interpretation' generally refers to an assortment of extra-scientific statements that are intended to 'save the theory'. They are neither verifiable nor falsifiable and consequently have no role whatsoever in science. Somehow they hang in there as some sort of an assurance that everything is 'intuitive' after all and so serve to make some people feel happy. The theses of this treatise are all scientific and are subject to the standard critiques of scientific methodology. Any assertion that I make may be right, or it may be wrong. However, in the contemporary sense of the word I am not giving a new 'interpretation' of quantum theory, but rather a new axiomatic approach to the already experimentally well established quantum theory. So, I consider this treatise to be a new axiomatization of standard quantum theory and in no way an 'interpretation' of it. Still, some comments are called for about some of the more common aspects of 'interpretations' of quantum theory. These 'interpretations' are numerous and often quite complicated, thereby making a case-by-case analysis tedious. Instead, I shall constrain myself to some general comments.

The reason these 'interpretations' are so prevalent is that they try to save a version of quantum theory which has neither consecutive nor conditional probability and which is inadequate for dealing with multiple events. This is why explaining 'collapse' is a frequent concern. So, all such 'interpretations' are rendered superfluous by the axiomatic approach of this treatise.

Others are concerned with the origin of probability in quantum theory. So, attempts are made to derive Born's Rule from more basic — perhaps, more intuitive — principles. But such attempts which only derive the single event Born's Rule are inadequate. And, as far as I am aware, there is no research on ways to derive Born's Rule for multiple events. Especially important would be the derivation of consecutive probability which, of course, depends on the time order of the events and so is a non-commutative probability.

Another sort of 'interpretation' deals with the relation between quantum systems and classical systems, particularly with classical measuring devices. Strictly speaking, this is not a problem just about quantum theory. However, this leads to theoretical measurement problems and therefore to theoretical solutions. One approach is to incorporate the seemingly classical measuring devices into the realm of quantum theory, while another is to explain the relation between quantum systems and purely classical measuring devices. These 'interpretations' can include extensive theories about measurement as a special sort of physical process. However, in the approach of this treatise measurements are just a type of quantum event and nothing more.

Again, as far as I am aware, none of the major 'interpretations' deals with consecutive and conditional probability of multiple events. So, they do not even consider saving a quantum theory that saves the phenomena of multiple events. This makes such 'interpretations' quite irrelevant.

Of course, since any theory should save the appropriate phenomena, there must be a clear agreement of how the theory does that. Such an agreement might well be labeled an 'interpretation' by some, but I prefer to avoid this now poisoned word when discussing quantum theory.

The quantum theory of this treatise is not designed, as it seems some 'interpretations' have been, in order to be intuitive or to make anyone happy, though that may happen too. It is meant to *save the phenomena* of the relative frequencies measured in experiments.

I include this chapter only to distance the contents of this treatise from all the ado that is called 'interpretation'.

Chapter 11

The Wave Function

> When we want to understand something strange,
> previously unknown to anyone, we have to begin
> with an entirely different set of questions.
> What is it? How does it work?
>
> — Margaret Mead

I do not want to discuss classical physics at all. But the wave function has the dubious role of being an aspect of quantum theory that is often considered in language that is classical. Here is some of that language. In classical physics the time evolution is described in terms of the points in a particular space, which is called the *phase space*. These points are the classical pure states. The phase space typically has high dimension. Then the classical dynamics specifies the possible one-dimensional curves in phase space. These curves are parameterized by time. Each point on such a curve represents a complete description of the physical system at the corresponding time.

All of this is temptingly analogous to the Schrödinger model of quantum theory, where the time dependent solution of Schrödinger's equation, which is inappropriately[1] called a *wave function*, gives a curve in a space whose points are called pure states. But analogies are not explanations. Unfortunately, this one is completely misleading, since in the isomorphic Heisenberg model the state is always constant in time. In classical physics the dynamical curve wending its way through a high dimensional phase space is used to deduce how physical objects move in three dimensional physical space. Analogously, the time dependent wave function is used to deduce properties that one can visualize in terms of the geometry of three dimensional physical space. However, the analogy fails in part because

---

[1] Schrödinger's equation is not a wave equation.

Schrödinger's equation is not the fundamental time evolution equation of quantum theory.

One way classical thinking can creep into quantum theory is when the Hilbert space for the quantum theory is $L^2(\mathbb{R}^3)$, since a normalized wave function ψ in that space has an associated probability density $|\psi|^2$ on $\mathbb{R}^3$. So, computer displays of these densities are made and can be seen in textbooks, in popular expositions and on the Internet. The graph of $|\psi|^2 : \mathbb{R}^3 \to [0, \infty)$ is a subset of $\mathbb{R}^4$, making direct visualization a bit tricky for most of us. But level sets in $\mathbb{R}^3$ can be encoded in computer memory and then their 2-dimensional projections can be displayed on a screen or printed on paper. If one uses instead a curve of normalized solutions of the time dependent Schrödinger's equation, then one can make a computer video. Of course, it is easy to misinterpret what the moving blob in such a video means. But a more profound problem is that, except for toy models, $L^2(\mathbb{R}^3)$ is not the Hilbert space used to describe quantum systems. In the following examples spin and statistics are omitted, since even without that the basic idea is clear.

Consider the ever popular example of the hydrogen atom. In the texts the stationary states $\psi \in L^2(\mathbb{R}^3)$ are found, together with their corresponding energy levels. Many images have been made for $|\psi|^2$. The formulas for the stationary states as well as the images made from them are all quite pretty. But all of that is misleading, since the (electrically neutral) hydrogen atom is a two body problem, whose Hilbert space in the Schrödinger model is $L^2(\mathbb{R}^6)$ with Euclidean coordinates x_1, x_2, x_3 for the electron and y_1, y_2, y_3 for the nucleus. A mathematical technique (change to center-of-mass coordinates) shows that the Hamiltonian of this system can be transformed in such a way as to give an equivalent problem with two Schrödinger operators, each acting in $L^2(\mathbb{R}^3)$. But neither one of these operators is the Hamiltonian of the hydrogen atom nor of any other physical entity. The wave functions for each of these two Schrödinger operators can be combined and then with an inverse change of coordinates can be written as $\psi(x_1, x_2, x_3, y_1, y_2, y_3)$ in $L^2(\mathbb{R}^6)$. But this last step is rarely done in the elementary texts. Rather only one of these two Schrödinger operators acting in $L^2(\mathbb{R}^3)$ is analyzed. For example this is what I did in [39].

Sometimes, instead of presenting the change of coordinates, a model is used with the nucleus fixed and immovable at some point in space. Then the electron is analyzed with a one-body Schrödinger operator. Among other things the mass parameter in such an approach should not be the electron mass. In fact, the mass parameter to be used depends on the isotope of the nucleus. But justifying that is difficult, to say the least. So, the electron mass is used, which introduces a small error. But when the electron-positron two body problem is considered, the error is no longer small. Worse yet, this 'approximation' contradicts the supposedly sacred Uncertainty Principle by fixing the values of both the position and the momentum of the nucleus.

The neutral hydrogen molecule is a four body problem and so has its wave functions in $L^2(\mathbb{R}^{12})$. The water molecule is a 21 body problem and so has its wave function in $L^2(\mathbb{R}^{63})$. And so it goes. But you can find images of the wave function of the ground state of the water molecule H_2O in the literature. This is done by changing to center-of-mass coordinates, fixing the values all of the variables for the 3 nuclei as well as integrating out all the variables for 17 of the 18 electrons. The modulus squared of the resulting 'wave function' in $L^2(\mathbb{R}^3)$ can then be published in a chemistry textbook or on the cover of a popular science magazine. This sort of visualization loses a lot of information that is encoded in the correct wave function of the molecule. Of course, it is approximately correct information that remains. (It is only approximate since there are only approximations of the original wave function itself.) But one is easily mislead into thinking that one has visualized the spatial structure of the molecule.

I will certainly be criticized for downplaying the role of Schrödinger's equation in quantum theory. But that equation stands firm on the basis of its many successes, which are found in the scientific literature. I emphasize in [39] the central role of Schrödinger's equation in the scientific activity of quantum physics. The two-slit experiment is but one grain of sand on the vast beach of successes of the Schrödinger model. They all speak for themselves[2]. I need not repeat those details. I only wish to place things in the correct perspective with respect to the Generalized Born's Rule. Those who wish to continue working with the Schrödinger model do not need permission to do so. Their results, if obtained correctly, will have scientific value. However, if one gives a physical significance to the model dependent aspects of the Schrödinger model, then I claim one is *overinterpreting* the mathematics.

I most likely expect to be criticized for taking a viewpoint that is 'too mathematical' in this treatise. But ironically it is those persons ascribing

---

[2]In poker the saying is that the cards speak for themselves. However, if the two-slit experiment does not speak clearly to you, see for example [3] for an explanation of it in terms of events, states and probabilities

deep meanings to mathematical structures with no physical significance who are the ones that are 'too mathematical'.

This topic is related to *quantization*, even though that is not a part of quantum theory but rather a way of arriving at a quantum theory from some other starting point. Typically, one starts from classical physics, but that is not essential. Actually, *second quantization* starts with one quantum theory in order to produce another. I do not wish to belittle research on quantization. It has its importance, but only as a way to quantum theory. In this treatise I want to describe the basics of the quantum world, and not go into the details of the journey for getting there. I have even done research on quantization, but this is not the place to go into that. My only goal now is to comment on its lack of relevance to understanding quantum theory in and of itself. Even worse, quantization is used to leak ideas of classical physics into quantum theory. Of course, some classical ideas do transfer to quantum theory, but others not. So, that issue must be faced and understood on a case by case basis.

Chapter 12

A Proposal and a Farewell

Bidding good-bye is such sweet sorrow
that I could bid adieu til it be morrow
Romeo in *Romeo and Juliette*

— William Shakespeare

It seems to me that the expression 'quantum mechanics' does not adequately describe the physical theory. It is based on two misleading words. The first is 'mechanics' which originally meant the study of machines but changed to mean the study of motion. Unfortunately, the concept of motion (as understood as an object moving along a curve) is not relevant to the physics we wish to discuss at the atomic and smaller length scales.

The word 'quantum' is even more unfortunate. It was used to contrast the new theory with classical theory where all measured values could assume a continuum of possible values. All of a sudden there were physical quantities whose measured values were in a discrete set. These measured values were said to be 'quantized'. While this is an important part of the story, it is not the whole story. Even in this new theory there are still some physical quantities with values in a continuum, such as the energy of a free particle. Other quantities are always 'quantized', such as the energy of a harmonic oscillator (realized as a diatomic molecule, say). But there are mixed cases, such as the hydrogen atom (with fixed nucleus), which has 'quantized' energy levels as well as a continuum of energy levels.

What is important here are the probabilities that arise from the pvm of a self-adjoint operator and more generally its spectral theory. So, it seems to be more appropriate to say *Spectral Probability Physics* instead of saying Quantum Mechanics for the physical theory based on self-adjoint operators. Another possibility is *von Neumann Algebra Probability Physics*.

Whatever one wishes to call it, this is an ongoing scientific project, whether it is done in the language of the Schrödinger model (with Schrödinger's equation and collapse), in the Heisenberg model (with its own mathematical tools) or any other model satisfying the Axioms, but always including the Generalized Born's Rule.

With this modest proposal I bid my gentle reader good-bye.

Bibliography

Books must follow sciences,
and not sciences books.

— Francis Bacon

[1] L. Accardi, Y.G. Lu and I. Volovich, *Quantum Theory and Its Stochastic Limit*, Springer, Berlin, 2002.

[2] V. Bargmann, On a Hilbert space of analytic functions and an associated integral transform, I, *Commun. Pure Appl. Math.* **14** (1961), 187–214.

[3] E.G. Beltrametti and G. Cassinelli, *The Logic of Quantum Mechanics*, Addison-Wesley, Reading, Mass, 1981.

[4] I. Bengtsson and K. Zyczkowski, *The Geometry of Quantum States, An Introduction to Quantum Entanglement*, Cambridge University Press, Cambridge, 2006.

[5] M. Born, Zur Quantemmechanik der Stossvorgänge, *Z. Physik*, **37** (1926), 863–867.

[6] O. Bratteli and D.W. Robertson, *Operator Algebras and Quantum Statistical Mechanics 1: C^*- and W^*-Algebras. Symmetry Groups. Decomposition of States*, 2nd. Ed., Springer, Berlin, 1987.

[7] P. Busch *et al.*, *Quantum Measurement*, Springer, Cham, 2016.

[8] G. Cassinelli and N. Zanghi, Conditional probabilities in quantum mechanics, *Nuovo Cimento* **73**(B) (1983), 237–245.

[9] G. Chiribella and R.W. Spekken (eds.), *Quantum Theory: Informational Foundations and Foils*, Springer, Dordrecht, 2016.

[10] E.B. Davies, Quantum stochastic processes, *Commun. Math. Phys.* **15** (1969), 277–304.

[11] E.B. Davies, Quantum stochastic processes II, *Commun. Math. Phys.* **19** (1970), 83–105.

[12] E.B. Davies, Quantum stochastic processes III, *Commun. Math. Phys.* **22** (1971), 51–70.

[13] D. Deutsch and C. Marletto, Constructor theory of information, *Proc. Royal Soc. A*, **471** (2015), 20140540.

[14] P.A.M. Dirac, *Principles of Quantum Mechanics*, 4th Ed., Oxford, 1958.

[15] A. Einstein, B. Podolsky and N. Rosen, Can quantum-mechanical description of physical reality be considered complete?, *Phys. Rev.* **47** (1935), 777–780.

[16] J. Frohlich and A. Pizzo, The time-evolution of states in quantum mechanics according to the *ETH*-approach, *Commun. Math. Phys.* **389** (2022), 1673–1715.

[17] A.M. Gleason, Measures on the closed subspaces of a Hilbert space, *J. Math. Mech.* **6** (1957), 885–893.

[18] R. Haag, *Local Quantum Physics*, 2nd ed., Springer, Berlin, 1996.

[19] R. Hudson, A short walk in quantum probability, *Phil. Trans. R. Soc.* A **376** (2018), 20170226.

[20] R.S. Ingarden and A. Kossakowski, On the connection of nonequilibrium information thermodynamics with non-Hamiltonian quantum mechanics of open systems, *Ann. Phys.* **89** (1975), 451–485.

[21] C. Isham, *Lectures on Quantum Theory: Mathematical and Structural Foundations*, Imperial College Press, London, 1995.

[22] T. Iwai, *Geometry, Mechanics, and Control in Action for the Falling Cat*, Springer, Singapore, 2021.

[23] R.V. Kadison and I. Singer, Extensions of pure states. *Am. J. Math.* **81** (1959), 383–400.

[24] R.V. Kadison, *Fundamentals of the Theory of Operator Algebras*, Vol. I, Academic Press, New York, 1983.

[25] A.N. Kolmogorov, *Foundations of the Theory of Probability*, 2nd ed., Chelsea Pub. Co., New York, 1956. (English translation of: *Grundbegriffe der Wahrscheinlichkeitsrechnung*, Springer, Berlin, 1933.)

[26] A. Kossakowski, On quantum statistical mechanics of non-Hamiltonian systems, *Rep. Math. Phys.* **3** (1972), 247–274.

[27] G. Lindblad, On the generators of quantum dynamical semigroups, *Commun. Math. Phys.* **48** (1976), 119–130.

[28] G. Lüders, Über die Zustandsänderung durch Messprozess, *Ann. Phys.* (Leipzig), **8** (1951), 322–328. English translation: Concerning the state-change due to the measurement process, *Ann. Phys.* (Leipzig) **15** (2006), 663–670.

[29] A. Marcus, D. Spielman and N. Srivastava, Interlacing families II: mixed characteristic polynomials and the Kadison-Singer problem. *Ann. Math.* **182** (2015), 327–350.

[30] J. Mehra and E.C.G. Sudarshan, Some reflections on the nature of entropy, irreversibility and the second law of thermodynamics, *Nuovo Cimento*, **11B** (1972), 215–256.

[31] N.D. Mermin, *Quantum Computer Science*, Cambridge University Press, Cambridge, 2007.

[32] J.E. Moyal, Quantum mechanics as a statistical theory, *Math. Proc. Cambridge Phil. Soc.*, **45** (1949), 99–124.

[33] J. Park, The concept of transition in quantum mechanics, *Found. Phys.* **1** (1970), 23–33.

[34] K.R. Parthasarathy, *An Introduction to Quantum Stochastic Calculus*, Monographs in Math. Vol. 85, Birkhäuser Verlag, Basel, 1992.

[35] L. Russo, *The Forgotten Revolution*, Springer, Berlin, 2004.

[36] E. Schrödinger, Quantisierung als Eigenwertproblem (Erste Mitteilung), *Ann. Phys.* **79** (1926), 361–376.

[37] E. Schrödinger, Discussion of probability relations between separate systems, *Proc. Camb. Phil. Soc.*, **31** (1935), 55.

[38] K.B. Sinha and D. Goswami, *Quantum Stochastic Processes and Noncommutative Geometry*, Cambridge University Press, Cambridge, 2007.

[39] S.B. Sontz, *An Introductory Path to Quantum Theory*, Springer, Cham, 2020.

[40] S.B. Sontz, A new organization of quantum theory based on quantum probability, *Found. Phys.* **53** (2023), 49. DOI:10.1007/s10701-023-00691-0.

[41] J. von Neumann, *Mathematical Foundations of Quantum Mechanics*, Princeton Univ. Press, Princeton, 1955. (English translation of: *Mathematische Grundlagen der Quantenmechanik*, Springer, Berlin, 1932.)

[42] J. Weidmann, *Linear Operators in Hilbert Spaces*, Springer, New York, 1980. (English translation of: *Lineare Operatoren in Hilberträumen*, B.G. Teubner, 1976.)

[43] J.A. Wheeler, Information, physics, quantum: Search for links, in: W.H. Zurek (ed.), *Complexity, Entropy, and the Physics of Information*, Addison-Wesley, Redwood City, Calif, (1990).

[44] E.P. Wigner, The problem of measurement, *Am. J. Phys.* **31** (1963), 6–15.

[45] P. Woit, *Quantum Theory, Groups and Representations*, Springer, Cham, 2017.

Index

A

Accardi, Luigi, 24
affiliated, 3, 8
algebra
  C*, 3–4, 12, 36, 65, 133
  linear, 25, 29–30
  von Neumann, 3–4, 6, 8–9, 12, 16,
    19, 21, 44, 61, 75, 91, 96–99,
    109–110, 122–123, 131
algorithm, 50–52, 89, 103, 122
Amis, Kingsley, 95
Aristotle, 62, 117
axioms, 1–23, 35, 41, 52–53, 56, 62,
  68, 84, 88, 98, 101–102, 109,
  118

B

Bachelard, Gaston, 107, 121
Bacon, Francis, 133
Bayes' Theorem, 36, 42–43
  classical, 42
  quantum, 43
beta decay, 4, 100, 104–105, 116
Borel
  σ-algebra, 5, 26–27, 48, 53, 62, 65
  function, 16, 28, 64
  measurable, 27, 63–64
  set, 10, 26–28, 40–42, 61–64
Born, Max, 33
Born's Rule
  generalized, 2, 14, 20, 24, 38–40,
    43, 53–54, 60, 66–68, 70–71, 77,
    91, 103, 111, 122, 129

time dependent, 13–14, 35
time independent, 10–12
boson, 8
bra, 15

C

Carroll, Lewis, vii
cause, 117–118
chance, 113, 119
classical
  mechanics, 89, 100, 114
  physics, 6, 42, 89, 96, 100, 103,
    105, 113, 127, 130
collapse, 20, 37–39, 50–52, 70, 76–77,
  79, 88–91, 102–103, 115, 122, 126,
  132
commutative, 3, 5–6, 9, 12, 27, 44, 47,
  65, 97–98
commutator, 15
complementarity, 22, 47, 62, 65
complete description, 25, 73, 107–108,
  127
confusion, 51, 67, 69, 74, 99, 109, 113,
  121, 123
conservation
  of probability, 13–14, 17, 19, 42,
    110
continuity, 7, 18, 116
correlation, 35, 38–39, 45, 57, 66, 71,
  74, 109, 121–123
covariance, 14, 17, 38

D

dark matter, 8
decoherence, 56
de Morgan, 62–63
density matrix, 3, 7, 10, 12–13, 17–19,
    33, 43–44, 48–50, 52, 54, 59–60, 64,
    68, 75
determinism, 79, 89, 113–115
diagonalization, 2, 30–31, 64
differential equation, 14–16, 69–70,
    73, 76, 108, 113, 115
Dirac notation, 3, 12, 33, 81
distribution, 63
dynamics, 12–13, 67, 102, 127, 134

E

effect, 89, 114, 117–118
eigenstate, 21, 93, 99
eigenvalue, 2, 4, 21, 30–31, 61–62, 81,
    98, 110
eigenvalue problem, 61–62
eigenvector, 2, 30, 61, 98, 102
Einstein, Albert, 91, 107
energy, 61, 101, 128, 131
entanglement, 6, 9, 45–46, 53, 56–57,
    79–97, 103, 107, 109, 115, 122–123
entropy, 71, 134–135
EPR paper, 73, 80, 107, 109–111, 119
equation
    auxiliary, 69
    differential, 14–16, 69–70, 73, 76,
        108, 113–115
    Dirac, 35
    Heisenberg, 16
    Klein–Gordon, 35
    Schrödinger, 21, 128
essential range, 63
Euclidean geometry, 101, 118–119
events
    empty sequence of, 52, 54, 56
    entangled, 85–86, 95
    independent, 45
    lattice of, 6
    multiple, 10, 35–36, 38, 73–74,
        122–123, 126

physical, 3–4, 8, 10, 12, 24, 39, 51,
    71–73, 91, 111, 121
quantum, 4–5, 8–10, 12–13, 16–17,
    26, 28, 34, 37, 39, 44–46, 51,
    58–59, 63–64, 71, 73, 81, 86,
    90–91, 96–98, 104–105, 109, 122,
    126
sequence of, 48, 52–56, 60, 68, 94
single, 10, 23, 35–36, 38–39, 51, 57,
    66, 74–75, 121–123, 126
theoretical, 23, 71
expectation, 9, 59–60, 69
expected value, 10, 34, 63–64, 66–67

F

fermion, 8
France, Anatole, 23
frequency, relative, 12, 39, 72, 97,
    101, 121
function
    Borel, 16, 28, 64
    bounded, 28
    characteristic, 31, 61, 63, 65
    essentially bounded, 28
    measurable, 64
functional calculus, 64, 115

G

Generalized Born's Rule, 2, 14, 20,
    23–24, 35, 38–40, 43, 52–56, 60,
    66–68, 70–71, 76, 103, 111,
    121–122, 132
graviton, 8
group
    Lie, 8, 70
    one-parameter, 12–14, 18–19, 53

H

Haag, Rudolf, 96–97
Hamiltonian, 9, 14, 60–61, 68–69, 76,
    102, 114, 128
Heisenberg model, 15–20, 38, 53, 67,
    70, 73, 76, 89, 104, 114–115, 127,
    132
Hermitian matrix, 29, 31

Hilbert space, 1–3, 5–6, 8–9, 14, 18–19, 21, 24–26, 28–29, 32–34, 47–48, 58–62, 69, 71, 76–77, 80–81, 85–87, 91, 94, 96, 98, 109, 114–115, 122, 128, 133–135
hydrogen atom, 128, 131

I

independence, 37, 45–46, 48, 53, 57, 80, 86, 93–94, 122
information, 2, 67, 71, 79, 83, 96, 100, 105, 110, 113, 122–123, 129
inner product, 3, 25, 58, 77, 80
instantaneous, 38, 51
integral
  non-commutative, 34
  time ordered, 59
interference terms, 42, 59, 66
interpretation, 21, 23, 38, 77, 90, 125–126
intuition, 16–18, 54, 57, 66, 84–85, 122
invariance
  Galilean, 41
  Lorentz, 41
  unitary, 40
Isham, Chris, 121
isomorphic, 18–19, 21, 38, 40, 69, 76, 85, 98, 127
isomorphism
  spatial, 21
  unitary, 21

J

Johnson, Samuel, 86

K

Kadison–Singer conjecture, 134
ket, 3, 15
kinematics, 3, 13
Kolmogorov, Andrei Nikolayevich, 23–24, 46, 62, 113, 115
Kronecker delta, 30, 88

L

lattice, 6, 63, 65
Lie group, 8
logic
  Boolean, 72
  classical, 62–63
  quantum, 72
Lüders Rule, 49–51

M

marginals, 43, 46, 60
mass, 116, 128–129
Maxwell's equations, 76
Mead, Margaret, 127
measure
  Lebesgue, 27
  positive operator valued, 65
  probability, 10, 12–14, 17, 19, 23, 32–37, 43, 63–67, 115–116
  projection valued, 3, 5, 9
  theory, 10, 26–28, 31–32, 60, 62–63
measurement, 2, 4, 9, 22, 25, 28, 32–33, 35, 47, 56, 74, 90, 100–105
measurement problem, 47, 99, 101–105, 122–123, 126
model
  Heisenberg, 15–20, 38, 53, 67, 70, 73, 76, 89, 104, 114–115, 127, 132
  interaction, 15, 18, 56, 60, 69, 73
  Schrödinger, 14–15, 17–21, 35, 38, 68–69, 73, 76, 89–91, 103, 114–115, 127–129, 132
moment
  central, 67
  first, 66–67
  kth, 66
motion, 20, 70, 100, 113, 131

N

Newton's Laws of Motion, 113
non-local, 38, 51, 79, 83
normalization, 7, 10, 32–34, 44, 46, 54–55, 60

O

observable, 1, 3–5, 10, 13–17, 34, 45,
  53–54, 58, 61–65, 67, 74, 80, 98–99
operator
  anti-linear, 3
  bounded, 15, 26, 29, 68
  entangled, 53, 79–81
  identically distributed, 54
  independent, 45–46, 53
  linear, 25–26, 29, 115
  position, 41, 58, 61
  positive, 50, 65
  projection, 4–5, 28, 32, 48, 123
  self-adjoint, 2–5, 8–10, 12–13,
    15–19, 21, 25, 28–29, 32–33, 36,
    39–42, 45, 53–54, 56, 62, 64–65,
    74, 80–81, 97, 99, 122, 131
  time evolution, 15, 68–69
  trace class, 7, 10–11, 33, 49
  unitary, 6, 18, 115
operator norm, 3, 7, 26–27

P

paradox, paradoxical, 38, 111, 116
partial trace, 9, 60, 71, 88
Pascal, Blaise, 117
phenomena, 1, 21–22, 34, 56, 70, 76,
  89–90, 98–99, 113, 118–119, 125
philosophical questions, 117, 119
physical reality, 73, 107–111, 119
physical significance, 8, 12, 14, 16, 18,
  21, 35, 39, 57–58, 61–62, 69, 81,
  83–84, 90–91, 98, 125, 129–130
picture
  Heisenberg, 15
  interaction, 18
  Schrödinger, 13–14
Planck's constant, xvi
positivity preservation, 7
probability
  classical, 23–25, 32–38, 43, 48, 54,
    59, 62–66
  conditional, 9, 19, 35–40, 49–55,
    57, 60, 66, 68, 70–71, 74–77,
    81–84, 86, 89–90, 103–105

consecutive, 39–41, 45–48, 52, 68,
  72–73, 92–93, 126
conservation of, 13–14, 17, 19
density, 39
non-commutative, 12, 24, 126
preservation of, 19
quantum, 10, 12, 23–77, 80, 90,
  115, 121–122
theory, 12, 23–25, 32–33, 36–38, 46,
  48, 59, 64–66, 72, 79, 83, 89, 110,
  113, 115, 119, 122
projection postulate, 38
projection valued measure (pvm), 3,
  5, 9

Q

quantization, 6, 8, 130
quantum
  σ-additivity, 10, 27, 44,
    46–47
  conditional expectation, 60
  event, 4–5, 8–10, 12–13, 16, 26, 28,
    34, 37, 39, 44–46, 51, 58, 63–64,
    71, 73, 81, 86, 91, 96–98,
    104–105, 109, 122, 126
  field theory, 9, 55
  integral, 28–29, 31, 59, 64, 66
  jump, 38
  mechanics, 1, 70, 77, 79, 108, 131,
    133–135
  probability, 10, 12, 23–77, 80, 90,
    115, 121–122
  state, 90
  system, 3, 6, 8–9, 24–25, 33, 68, 73,
    80, 88, 90, 99–100, 103, 105, 109,
    114, 123, 126, 128

R

randomness, 119
random variable, 53, 63
reduction of state vector, 121–122
resolution of the identity, 30
resolvent set, 29
Roberts inequality, 67
Russell, Bertrand, 1

S

save the phenomena, 89, 98, 125–126
Schrödinger equation, 61, 73, 108
Schrödinger, Erwin, 79
Schrödinger model, 14–15, 17–18, 20,
38, 68–69, 73, 76, 89–91, 103,
114–115, 127–129, 132
Schrödinger's razor, 76
science, 8, 24, 35, 38–39, 57, 71, 74,
76, 91, 97, 119, 121–122, 125, 129
self-interference, 47
semi-group, 134
quantum dynamical, 134
Shakespeare, William, 73, 101, 131
σ-additivity, 10, 27, 44, 46–47
Sontag, Susan, 125
space
Banach, 7, 26
configuration, 1, 3, 5–6, 8–9, 14,
18–19, 24–26, 28–29, 32–34,
47–48, 58–59, 61–62, 77, 80–81,
85–86, 91, 94, 96, 98, 109, 115,
128
Minkowski, 95
phase, 100, 127
probability, 35
projective, 8
sample, 62
Segal–Bargmann, 21
vector, 27–28, 63
space-like, 41, 79–80, 89, 96, 109, 116
spacetime, 38, 51, 71
spectral measure, 33, 65
spectral representation, 31
spectral subspace, 30
spectral theorem, 2, 11, 15, 25, 28–29,
31–32, 64
spectrum, 4–5, 17, 29, 63–64, 74, 81,
96–97
spin, 8–10, 41, 80–84, 89, 91, 93–94,
98–99, 110, 116, 128
spin matrices, 80
standard deviation, 67
state
entangled, 48, 86–88

ground, 129
mixed, 7, 13, 25
pure, 7, 9, 11, 13–16, 18, 21, 25, 31,
43, 49, 74–75, 87, 92–93, 98, 109,
123, 127
quantum, 90, 133
statistics, 8–9, 128
Stone's theorem, 16
superposition, 47, 98
superselection rule, 25, 98
support, 28, 74

T

tensor product, 45, 48, 54, 60, 71,
86–87, 91, 94
time evolution, 10, 13–17, 20–21,
23–24, 35, 53–54, 60, 62, 65–70, 76,
91, 102–103, 108, 115, 121,
127–128
time ordered product, 39, 45, 52, 59
topology
metric, 27
norm, 27
operator norm, 27
strong operator, 11, 27–28, 116
trace, 7, 9–11, 33–34, 49–50, 60, 71,
88
type I error, 6, 98

U

uncertainty principle, 47, 67,
129
unitarity, 14
unitary group, 5, 16, 18, 68
unitary representation, 5, 18, 21
unit vector, 3, 7–8, 13–14, 21, 24, 33,
36, 39, 43, 45–46, 48, 51–52, 57–59,
61, 66, 70, 74–75, 81–82, 86–87,
92–93, 98

V

variance, 67
von Neumann, John, 3–4, 6, 8–9, 12,
16, 19, 21, 23, 44, 61, 67, 75, 91,
96–99, 109–110, 123, 131

W

Wallace, Alfred Russel, 35
wave function, 20, 38, 73, 107–109,
    111, 127–129
Weyl–Heisenberg group, 10
Wigner's friend, 99
Wigner's Rule, 52, 68

Wigner's theorem, 39
Wigner transformation,
    39
world-line, 95–96

Y

Yes-No experiment, 17

Series on the Foundations of Natural Science and Technology

(Continued from page ii)

Vol. 9   *The Timeless Approach: Frontier Perspectives in 21st Century Physics*
          by D. Fiscaletti

Vol. 8   *Quantum Effects, Heavy Doping, and the Effective Mass*
          by K. P. Ghatak

Vol. 7   *Magneto Thermoelectric Power in Heavily Doped Quantized Structures*
          by K. P. Ghatak

Vol. 6   *Nano-Engineering in Science and Technology: An Introduction
          to the World of Nano-Design*
          by M. Rieth

Vol. 5   *Grasping Reality: An Interpretation-Realistic Epistemology*
          by H. Lenk

Vol. 4   *What is Life? Scientific Approaches and Philosophical Positions*
          by H.-P. Dürr, F.-A. Popp and W. Schommers

Vol. 3   *The Visible and the Invisible: Matter and Mind in Physics*
          by W. Schommers

Vol. 2   *Symbols, Pictures and Quantum Reality: On the Theoretical
          Foundations of the Physical Universe*
          by W. Schommers

Vol. 1   *Space and Time, Matter and Mind: The Relationship between
          Reality and Space-Time*
          by W. Schommers

Forthcoming:

Vol. 10   *Space-Time Reality: Mach's Principle, Relativity and Novel Aspects*
          by W. Schommers

Vol. 11   *The Origin of Mind: Interplay between Matter and Mind*
          by W. Schommers

www.ingramcontent.com/pod-product-compliance
Lightning Source LLC
Chambersburg PA
CBHW050642190326
41458CB00008B/2380